Luigi Palmieri, Robert Mallet

The eruption of Vesuvius in 1872

With notes and an introductory sketch of the present state of knowledge of

terrestrial vulcanicity

Luigi Palmieri, Robert Mallet

The eruption of Vesuvius in 1872
*With notes and an introductory sketch of the present state of knowledge of terrestrial
vulcanicity*

ISBN/EAN: 9783337257972

Printed in Europe, USA, Canada, Australia, Japan

Cover: Foto ©berggeist007 / pixelio.de

More available books at **www.hansebooks.com**

THE

ERUPTION OF VESUVIUS

IN 1872,

BY

PROFESSOR LUIGI PALMIERI,

Of the University of Naples ; Director of the Vesuvian Observatory.

WITH NOTES, AND AN

INTRODUCTORY SKETCH OF THE PRESENT STATE OF KNOWLEDGE

OF

TERRESTRIAL VULCANICITY,

The Cosmical Nature and Relations of

Volcanoes and Earthquakes.

BY

ROBERT MALLET,

Mem. Inst. C.E., F.R.S., F.G.S., M.R.I.A., &c., &c.

WITH ILLUSTRATIONS.

LONDON :

ASHER & CO.,

13, BEDFORD STREET, COVENT GARDEN, W.C.

1873.

"The Translator should look upon himself as a Merchant in the Intellectual Exchange of the world, whose business it is to promote the interchange of the produce of the mind."

GŒTHE, " *Kunst und Alterthum.*"

INTRODUCTORY SKETCH, &c.

THE publishers of this little volume, in requesting me to undertake a translation of the "Incendio Vesuviano," of Professor Palmieri, and to accompany it with some introductory remarks, have felt justified by the facts that Signor Palmieri's position as a physicist, the great advantages which his long residence in Naples as a Professor of the University, and for many years past Director of the Meteorological Observatory—established upon Vesuvius itself, prior to the expulsion of the late dynasty—have naturally caused much weight to attach to anything emanating from his pen in reference to that volcano.

Nearly forty memoirs on various branches of physics —chiefly electricity, magnetism and meteorology—produced since 1842, are to be found under Palmieri's name in the " Universal Catalogue of Scientific Papers of the Royal Society," and of these nine refer to Vesuvius, the earliest being entitled " Primi Studii Meteorologici fatti sul R. Osservatorio Vesuviano," published in 1853. He was also author, in conjunction with Professor A. Scacchi, of an elaborate report upon the Volcanic Region of Monte Vulture, and on the Earthquake (commonly called of Melfi) of 1851. These, however, by no means exhaust the stock of Palmieri's labours.

B

The following Memoir of Signor Palmieri on the eruption of Vesuvius in April of this year (1872), brief as it is, embraces two distinct subjects, viz., his narrative as an eye-witness of the actual events of the eruption as they occurred upon the cone and slopes of the mountain, and his observations as to pulses emanating from its interior, as indicated by his Seismograph, and as to the electric conditions of the overhanging cloud of smoke (so called) and ashes, as indicated by his bifilar electrometer, both established at the Observatory. The two last have but an indirect bearing upon Vulcanology. The narrative of the events of the eruption is characterised by exactness of observation and a sobriety of language—so widely different from the exaggerated style of sensational writing that is found in almost all such accounts—that I do the author no more than justice in thus expressing my view of its merits.

Nor should a special narration, such as this, become less important or suffer even in popular estimation by the fact that so recently my friend, Professor J. Phillips, has given to the world the best general account of Vesuvius, in its historical and some of its scientific aspects, which has yet appeared. That monograph— with its sparkling style, and scholarly digressions, as well as for its more direct merits—will, no doubt, become the manual for many a future visitor to the volcanic region of Naples; but it, like the following Memoir of Palmieri, and in common with almost every work that has appeared on the subject of Volcanoes, contains a good deal which, however interesting, and remotely related to Vulcanology, does not properly belong to the body of that branch of cosmical science, as I understand its nature and limits.

It tends but little, for example, to clear our views, or enlarge our knowledge of the vast mechanism in which the Volcano originates, and that by which its visible mass is formed, that we should ascertain the electric condition of the atmosphere above its eruptive cone, or into what crystallographic classes the mineral species found about it may be divided : it will help us but little to know Pliny's notions of how Pompeii was overwhelmed, or to re-engrave pictures, assumed to give the exact shape of the Vesuvian or other cone at different periods, or its precise altitude, which are ever varying, above the sea. Even much more time and labour may be spent upon analysing the vapours and gases of fumarolles and salfatares than the results can now justify.

Nothing, perhaps, tends more to the effective progress of any branch of observational and inductive science, than that we should endeavour to discern clearly the scope and boundary of our subject.

To do so is but to accord with Bacon's maxim, " *Prudens questio dimidium scientiæ.*" That once shaped, the roads or methods of approach become clearer ; and every foothold attained upon these direct paths enables us to look back upon such collateral or subordinate questions as at first perplexed us, and find them so illuminated that they are already probably solved, and, by solution, again prove to us that we *are* in the right paths.

I believe, therefore, that I shall not do disservice to

the grand portion of cosmical physics to which volcanic phenomena belong, by devoting the few pages accorded to me for this Introduction to sketching what seems to me to be the present position of terrestrial *Vulcanicity*, and tracing the outlines and relations of the two branches of scientific investigation—*Vulcanology* and *Seismology*—by which its true nature and part in the Cosmos are chiefly to be ascertained.

The general term, *Vulcanicity*, properly comprehends all that we see or know of actions taking place upon and modifying the surface of our globe, which are referable not to forces of origin above the surface, and acting superficially, but to causes that have been or are in operation beneath it. It embraces all that Humboldt has somewhat vaguely called "the reactions of the interior of a planet upon its exterior."

These reactions show themselves principally and mainly in the marking out and configuration of the great continents and ocean beds, in the forcing up of mountain chains, and in the varied phenomena consequent thereon, as seen in more or less adjacent formations.

These constitute the mechanism which has moulded and fashioned the surface of our globe from the period when it first became superficially solid, and prepared it as the theatre for the action of all those superficial actions—such as those of tides, waves, rain, rivers, solar heat, frost, vitality, vegetable and animal (passing by many others less obvious)—which perpetually modify, alter or renew the surface of our world, and maintain the existing regimen of the great machine, and of its inhabitants. These last are the domain of Geology, properly so called. No geological system can be well founded, or can completely explain the working of the world's system as we now see it, that does not start from Vulcanicity as thus defined; and this is equally true, whether, as do most geologists, we include within the term Geology everything we can know about our world as a whole, exclusive of what Astronomy teaches as to it, dividing Geology in general into Physical Geology—the boundaries of which are very indistinct—and Stratigraphical Geology, whose limits are equally so.

It has been often said that Geology in this widest sense begins where Astronomy or Cosmogony ends its information as to our globe, but this is scarcely true.

Vulcanicity—or Geology, if we choose to make it comprehend that—must commence its survey of our world as a nebula upon which, for unknown ages, thermic, gravitant and chemical forces were operative, and to the final play of which, the form, density and volume, as well as order of deposition of the different elements in the order of their chemical combination and deposition was due, when first our globe became a liquid or partly liquid spheroid, and which have equally determined the chemical nature of the materials of the outward rind of the earth that now is, and with these some of the primary conditions that have fixed the characters, nature and interdependence of the vegetables and animals that inhabit it. Physical Astronomy and Physical Geology, through Vulcanicity, thus overlap each other; the first does not end where the second begins; and in every sure attempt to bring Geology to that pinnacle which is the proper ideal of its completed design—namely, the interpretation of our world's machine, as part of the universal Cosmos (so far as that can ever become known to our limited observation and intelligence)—we must carry with us astronomic considerations, we must keep in view events anterior to the "*status consistentior*" of Leibnitz, nor lose sight of the fact that the chain of causation is one endless and unbroken; that forces first set moving, we know not when or how, the dim remoteness of which imagination tries to sound in shadowy thought, like those of the grand old Eastern poem, "When the morning stars first sang together," are, however changed in form, operative still. The light and fragile butterfly, whose glorious garb irradiates the summer zephyr in which it floats, has had its power of flight—which is its power to live—determined by results of that same chain of causes that lifted from the depths the mountain on whose sunny side he floats, that has determined the seasons and the colour of the flower whose nectar he sucks, and that discharges or dissipates the storm above, that may crush the insect and the blossom in which it basked. And thus, as has been said, it was not all a myth, that in older days affirmed that in some mysterious way the actions and the lives of men were linked to the stars in their courses.

Whatever may have been the manifestations of Vulcanicity at former and far remoter epochs of our planet, and to which I shall return, in the existing state of regimen of and upon our globe it shows itself chiefly in the phenomena of Volcanoes and of Earthquakes, which are the subjects of Vulcanology and of Seismology respectively, and in principal part, also, of this Introduction.

The phenomena of hot springs, geysers, etc., which might be included under the title of Thermopægology, have certain relations to both, but more immediately to Vulcanology.

Let us now glance at the history and progress of knowledge in these two chief domains of Vulcanicity, preparatory to a sketch of its existing stage as to both, and, by the way, attempt to extract a lesson as to the methods by which such success as has attended our labours has been achieved.

It will be most convenient to treat of Seismology first in order.

Aristotle—who devotes a larger space of his Fourth Book, Περὶ Κοσμου, to Earthquakes—Seneca, Pliny, Strabo, in the so-called classic days, and thence no end of writers down to about the end of the seventeenth century—amongst whom Fromondi (1527) and Travagini (1679) are, perhaps, the most important now—have filled volumes with records of facts, or what they took to be such, of Earthquakes, as handed down to or observed by themselves, and with plenty of hypotheses as to their nature and origin, but sterile of much real knowledge.

Hooke's "Discourses of Earthquakes," read before the Royal Society about 1690, afford a curious example of how abuse of words once given by authority clings as a hindrance to progress. He had formed no distinct idea of what he meant by an Earthquake, and so confusedly mixes up all elevations or depressions of a permanent character with "subversions, conversions and transpositions of parts of the earth," however sudden or transitory, under the name of Earthquakes.

A like confusion is far from uncommon amongst geological writers, even at the present day, and examples might be quoted from very late writings of even some of the great leaders of English Geology.

From the seventeenth to the middle of the eighteenth century one finds floods of hypotheses from Flamsteed, Höttinger, Amontons, Stukeley, Beccaria, Percival, Priestly, and a crowd of others, in which electricity, then attracting so much attention, is often called upon to supply causation for a something of which no clear idea had been formed. Count Bylandt's singular work, published in 1835, though showing a curious *partial* insight in point of advancement, might be put back into that preceding period.

In 1760 appeared the very remarkable Paper, in the fifty-first volume of the "Philosophical Transactions," of the Rev. John Mitchell, of Cambridge, in which he views an Earthquake as a sudden lifting up, by a rapid evolution of steam or gas beneath, of a portion of the earth's crust, and the lateral transfer of this gaseous bubble beneath the earth's crust, bent to follow its shape and motion, or that of a wave of liquid rock beneath, like a carpet shaken on air. Great as are certain collateral merits of Mitchell's Paper, showing observation of various sorts much in advance of his time, this notion of an Earthquake is such as, had he applied to it even the imperfect knowledge of mechanics and physics then possessed in 'a definite manner, he could scarcely have failed to see its untenable nature. That the same notion, and in a far more extravagant form, should have been reproduced in 1843 by Messrs. Rogers, by whom the gigantic parallel anticlinals, flanks and valleys of the whole Appalachian chain of mountains are taken for nothing more than the indurated foldings and wrinkles of Mitchell's carpet, is one of the most salient examples of the abuse of hypothesis untested by exact science.

Neither Humboldt nor Darwin, great as were the opportunities of observation enjoyed by both, can be supposed to have formed any definite idea of *what* an Earthquake is; and the latter, who had observed well the effects of great sea-waves rolling in-shore after the shock, did not establish any clear relation between the two.*

Hitherto no one appears to have formed any clear notion as to what an Earthquake is—that is to say, any clear idea of what is the nature of the movement constituting the shock, no matter what may be the nature or origin of the movement itself. The first glimmering of such an idea, so far as my reading has enabled me to ascertain, is due to the penetrating genius of Dr. Thomas Young, who, in his " Lectures on Natural Philosophy," published in 1807, casually suggests the probability that earthquake motions are vibratory, and

---

* For a fuller account of the literature and history of advancement of human knowledge as to Earthquakes, here merely glanced at, I must refer to my First Report on the Facts of Earthquakes, " Reports, British Association, 1850," and to the works of Daubeny, Lyall, Phillips and others, its *complete* history remaining yet to be written.

are analogous to those of sound.* This was rendered somewhat more definite by Gay Lussac, who, in an able paper " On the Chemical Theories of Volcanoes," in the twenty-second volume of the " Annales de Chémie," in 1823, says: " En un mot, les tremblements de terre ne sont que la propagation d'une commotion à travers la masse de la terre, tellement indépendante des cavités souterraines qu'elle s'entendrait, d'autant plus loin que la terre serait plus homogène."

These suggestions of Young and of Gay Lussac, as may be seen, only refer to the movement in the more or less solid crust of the earth. But two, if not three, other great movements were long known to frequently accompany earthquake shocks—the recession of the sea from the shore just about the moment of shock— the terrible sounds or subterraneous growlings which sometimes preceded, sometimes accompanied, and sometimes followed the shock—and the great sea-wave which rolls in-shore more or less long after it, remained still unknown as to their nature. They had been recognised only as concomitant but unconnected phenomena—the more inexplicable, because sometimes present, sometimes absent, and wholly without any known mutual bearing or community of cause.

On the 9th February, 1846, I communicated to the

---

* Yet how indistinctly formed were Young's ideas, and indistinct in the same direction as those of Humboldt, becomes evident by a single sentence : " When the agitation produced by an Earthquake extends further than there is any reason to suspect a subterraneous communication, it is probably propagated through the earth nearly in the same manner as a noise is conveyed through the air."—*Lectures, Nat. Phil.*, Vol. I.

Royal Irish Academy my Paper, "On the Dynamics of Earthquakes," printed in Vol. XXI., Part I., of the Transactions of that Academy, and published the same year in which it was my good fortune to have been able to colligate the observed facts, and bringing them together under the light of the known laws of production and propagation of vibratory waves in elastic, solid, liquid and gaseous bodies, and of the production and propagation of liquid waves of translation in water varying in depth, to prove that all the phenomena of earthquake shocks could be accounted for by a single impulse given at a single centre. The definition given by me in that Paper is that an earthquake is "*The transit of a wave or waves of elastic compression in any direction, from vertically upwards to horizontally, in any azimuth, through the crust and surface of the earth, from any centre of impulse or from more than one, and which may be attended with sound and tidal waves dependent upon the impulse and upon circumstances of position as to sea and land.*"

Thus, for example, if the impulse (whatever may be its cause) be delivered somewhere beneath the bed of the sea, all four classes of earthquake waves may reach an observer on shore in succession. The elastic wave of shock passing through the earth *generally* reaches him first: its velocity of propagation depending upon the specific elasticity and the degree of continuity of the rocky or the incoherent formations or materials through which it passes.

Under conditions pointed out by me, this elastic wave may cause an aqueous wave, producing reces-

sion of the sea, just as it reaches the margin of sea and
land.

If the impulse be attended by fractures of the
earth's crust, or other sufficient causes for the impulse to
be communicated to the air directly or through the
intervening sea, ordinary sound-waves will reach the
observer through the air, propagated at the rate of
1,140 feet per second, or thereabouts; and may also
reach him before or with or soon after the shock itself,
through the solid material of the earth; and lastly, if
the impulse be sufficient to disturb the sea-bottom
above the centre of impulse, or otherwise to generate an
aqueous wave of translation, that reaches the observer
last, rolling in-shore as the terrible " great sea-wave,"
which has ended so many of the great earthquakes, its
dimensions and its rate of propagation depending upon
the magnitude of the originating impulse and upon the
variable depth of the water.   It is not my purpose, nor
would it be possible within my limits here, to give any
complete account of the matter contained in that Paper,
which, in the words of the President of the Academy
upon a later occasion, " fixed upon an immutable basis
the true theory of Earthquakes."*   I should state, however, that in it I proved the fallacy of the notion of
vorticose shocks, which had been held from the days of
Aristotle, and showed that the effects (such as the
twisting on their bases of the Calabrian Obelisks) which

---

* The Right Rev. Charles Graves, F.R.S., etc., then Fellow of
Trinity College, Pres. R. I. Acad., and now Bishop of Limerick, on
presentation of the Academy's Cunningham Medal.

had been supposed due to such, were but resolved motions, due to the transit rectilinearly of the shock.

This removed one apparent stumbling block to the true theory.

Incidentally also it was shown that from the observed elements of the movement of the elastic wave of shock at certain points—by suitable instruments—the position and depth of the *focus*, or centre of impulse, might be inferred.

In the same volume ("Transactions of the Royal I. Academy," XXI.) I gave account, with a design to scale, for the first self-registering and recording seismometer ever, to my knowledge, proposed. In some respects in principle it resembles that of Professor Palmieri, of which he has made such extended use at the Vesuvian Observatory, though it differs much from the latter in detail. In June, 1847, Mr. Hopkins, of Cambridge, read his Report, "On the Geological Theories of Elevation and Earthquakes," to the British Association—requested by that body the year before—and printed in its Reports for that year.

The chief features of this document are a digest of Mr. Hopkins's previously published "Mathematical Papers" on the formations of fissures, etc., by elevations and depressions, and those on the thickness of the earth's crust, based on precession, etc., which he discusses in some relations to volcanic action.

This extends to forty-one pages, the remaining eighteen pages of the Report being devoted to "Vibratory Motions of the Earth's Crust produced by Subterranean Forces—Earthquakes."

The latter consists mainly of a *résumé* of the acknowledged laws, as delivered principally by Poisson, of
formation and propagation of elastic waves and of
liquid waves, by Webers, S. Russel and others—the
original matter in this Report is small—and as respects
the latter portion consists mainly in some problems for
finding analytically the position or depth of the centre
of disturbance when certain elements of the wave of
shock are given, or have been supposed registered by
seismometric instruments, such as that described by
myself, and above referred to.*　At the time my
original Paper " On the Dynamics of Earthquakes" was
published, there was little or no *experimental* knowledge as to the actual velocity of transit of waves—
analogous to those of sound, but of greater amplitude—

---

* In this Report, though I have never before referred to it, and
do so now with reluctance, I have always felt that the Author did
me some injustice.　The only reference made to my labours, published the preceding year only, is in the following words : " Many
persons have regarded these phenomena (viz., Earthquakes) as due in
a great measure to vibrations . . . and the subject has lately been
brought under our notice, in a Memoir by Mr. Mallet, ' On the
Dynamics of Earthquakes,' in which he has treated it in a more
determinate manner, and in more detail, than any preceding writer "
(p. 74).　If that Paper of mine be collated with this Report, it will
be, I believe, found that, as respects the earthquake part, the latter
tint parades, in a mathematical dress, some portions of the general
theory of earthquake movements, previously published by me as
above stated.　So, also, in the chapter (p. 90) referring to Seismometry, and the important uses to Geology that might be (and since
have been, to some extent) made of it, no mention is made of those
instruments previously proposed by me, nor of my anticipation of
their important uses.　This is but too mortifyingly suggestive of
the—

" Pereant qui mea ante mihi dixerunt."

Having left this unnoticed for so many years, and during which
the Author has preceded me to that bourne where our errors to each

through elastic solids. The velocity as deduced from theory, the solid being assumed quite *homogeneous* and *continuous*, was very great, and might be taken for some of the harder and denser rock formations at 11,000 or 12,000 feet per second. That these enormous velocities of wave transit would be something near those of actual earthquake shock seemed probable to me, and was so accepted by Hopkins.

Thus, he says (Report, p. 88): " The velocity of the sea-wave, for any probable depth of the sea, will be so small as compared with that of the vibratory wave, that we may consider the time of the arrival of the latter at the place of observation as coincident with that of the departure of the sea-wave from the centre of divergence."

---

other must be forgotten, I should certainly not have now trespassed on the good rule, *De mortuis nil nisi bonum*, had I not observed very recently one amongst other results probably attributable to it. In Professor Phillips's " Vesuvius," if any one will refer to the passage beginning "The mechanism of earthquake movement has been investigated by competent hands. The late eminent mathematician, Mr. Hopkins, explained these tremors in the solid earth by the general theory of vibratory motion," etc. (pages 257–259)—I think he must, in the absence of collateral information, conclude that, not I, but Mr. Hopkins, was the discoverer of the Theory of Earthquakes as explained by the general theory of vibratory motion.

Probably my friend, Professor Phillips, had not recently referred to those Memoirs and Reports of twenty-four years back, and I am thoroughly convinced that, if he has here perpetuated an injustice, he has done so unintentionally and unwittingly.

Still, the facts show how true it is that

" The ill men do lives after them,
The good they do is oft interred with their bones."

And I may venture to ask my friend, should his admirable book reach, as I doubt not it will, another edition, to modify the passage.

In my original Paper (Dynamic, &c.), I had suggested, as an important object, to ascertain by actual experiment what might be the wave's transit rate in various rocky and incoherent formations ; and having proposed this in my first " Report upon the Facts of Earthquake" to the British Association, I was enabled by its liberality to commence those experiments, in which I was ably assisted by my eldest son, then quite a lad—Dr. Jno. William Mallet, now Professor of Chemistry at the University of Virginia, U.S. ; and to give account of the results, in my second Report (" Report, British Association for 1851") to that body.

Those experiments were made by producing an impulse at one end of an accurately measured base line, by the explosion of gunpowder in the formation experimented upon, and noting the time the elastic wave generated required to pass over that distance, upon a nearly level surface. Special instruments were devised and employed, by which the powder was fired and the time registered, by touching a lever which completed certain galvanic contacts. The media or formations in which these experiments were conducted were, damp sand—as likely to give the minimum rate—and crystalline rock (granite), as likely to give the maximum. The results were received, not with doubt, but with much surprise, for it at once appeared that the actual velocity of transit was vastly below what theory had indicated as derivable from the density and modulus of elasticity of the material, taken as homogeneous, etc. The actual velocities in feet per second found were :

In sand . . . 824·915 feet per second.
In discontinuous and
   much   shattered
   granite . . 1,306·425 ,, ,,
In more solid granite 1,664·574 ,, ,,

This I at once attributed, and as it has since been proved correctly, to the loss of *vis viva*, and consequently of speed, by the *discontinuity of the materials*.

And some indication of the general truth of the fact was derivable from comparing the rude previous approximations to the transit rate of some great Earth-quakes. In the case of that of Lisbon, estimated by Mitchell at 1,760 feet per second. It was still desirable to extend similar experiments to the harder classes of stratified and of contorted rocks. This I was enabled to carry into effect, at the great Quarries at Holyhead (whence the slate and quartz rocks have been obtained for the construction of the Asylum Harbour there), taking advantage of the impulses generated at that period by the great mines of powder exploded in these rocks.

The results have been published in the "Philosophical Transactions for 1861 and 1862 (Appendix)." They show that the mean lowest rate of wave transit in those rocks, through measured ranges of from 5,038 to 6,582 feet, was 1,089 feet per second; and the mean highest, 1,352 feet per second; and the general mean 1,320 feet per second.

By a separate train of experiments on the com-pressibility of solid cubes of these rocks, I obtained the mean modulus of elasticity of the material when

c

perfectly continuous and unshattered, with this remarkable result—that in these rocks, as they exist at Holyhead,
*nearly seven-eighths of the full velocity of wave trans-
mission due to the material, if solid and continuous, is lost
by reason of the heterogeneity and discontinuity* of the
rocky masses as they are found piled together in
Nature.

I also proved that the wave-transit period of the
unshattered material of these rocks was greatest in a
direction *transverse* to the bedding, and least in line
parallel with that; but the effect of this in the rocky
mass itself may be *more* than counterbalanced by the
discontinuity and imperfect contact of the adjacent
beds.

These results indicate, therefore, that the superficial
rate of translation of the solitary sea-wave of earthquakes may, when over very deep water, equal or even
exceed the transit rate (in some cases) of the elastic
wave of shock itself.

These results have since received general confirmation
by the careful determinations of the transit rates of
actual earthquake waves, in the rocks of the Rhine
Country and in Hungary, by Nöggerath and Schmidt
respectively, and by those made since by myself in those
of Southern Italy, to which I shall again refer. In an
elastic wave propagated from a centre of impulse in an
infinitely extended volume of a perfect gas, normal
vibrations are alone propagated—as is the case with
sound in air.

In the case of like movements propagated in elastic
and perfectly homogeneous and isotropic solids, the

wave possesses both normal and transversal vibrations, and is, in so far, analogous to the case of light. Mr. Hopkins, in his Report above referred to, has based certain speculations upon the assumed necessary co-existence of both orders of vibration in actual earth-quake shocks in the materials of which our earthy crust is actually composed.

The existence of transversal vibration in those materials has not been yet proved experimentally, though there is sufficient ground to preclude our denying their probable existence.

That if they do exist they play but a very subordinate part in the observable phenomena of actual Earthquake is highly probable. This is the view, supported not only by observations of the effects of such shocks in Nature, but by the theoretic consideration of the effects of discontinuity of formations in planes or beds more or less transverse to the wave path (or line joining the centre of impulse with the mean centre of wave disturbance at any point of its transit). If we suppose, for illustration sake, such an elastic wave transmitted perpendicularly through a mass of glass plates, each indefinitely thin, and all in absolute contact with each other, but without adhesion or friction, more or less of the transversal vibration of the wave would be cut off and lost at each transit from plate to plate, as the elastic compression can, by the conditions, be transmitted only normally or by direct push perpendicularly from plate to plate. This must take place in Nature, and to a very great extent, and the consideration, with others, enabled me generally to apply the normal

wave motion of shock alone to my investigation as to
the depth of the centre of impulse of the great Neapolitan Earthquake of 1857, an account of which was
published in 1862, and to be presently further referred to.

Hitherto the multitudinous facts, or supposed facts,
recorded in numberless accounts of Earthquakes had
remained almost wholly unclassified, and so far as they
had been discussed—in a very partial manner, as inci-
dental portions of geological treatises—with little attempt
to sift the fabulous from the real, or to connect the
phenomena admitted by reference to any general mechanical or physical causes.   In 1850 my first " Report
upon the Facts of Earthquakes," called for by the
British Association in 1847, was read and published in
the Reports of that body for that year.   In this, for
the first time, the many recorded phenomena of Earthquakes are classified, and the important division of the
phenomena into primary and secondary effects of the
shock was established.   Several facts or phenomena,
previously held as marvellous or inexplicable, were
either, on sufficient grounds, rejected, or were, for the
first time, shown susceptible of explanation.   Amongst
the more noticeable results were the pointing out that
fissures and fractures of rock or of incoherent formations were but secondary effects, and, in the latter, were,
in fact, generally of the nature of inceptive landslips.
This last was not accepted, I believe, by geologists at
the time; but the correctness of the views then pro-
pounded as to earth fissures—the nature of the spouting
from them of water or mud—the appearances taken

for smoke issuing from them, etc.—have since been fully confirmed, first, by my own observations upon the effects of the Great Neapolitan Earthquake of 1857, and more lately by those of Dr. Oldham upon the Earthquake of Cachar (India), where he was enabled to observe fissures of immense magnitude, the nature of the production of which he has well described and explained in the "Proceedings, Geological Society, London, 1872."

The relations between meteorological phenomena proper and Earthquakes have always been a subject of popular belief and superstition.

This was here carefully discussed, and with the result of disproving any connection, or, if any, but of an indirect nature. I also, to some extent, towards the end of this Report, discussed the question of the possible nature of the *impulse itself* which originates the shock ; I showed that it must be of the nature of a blow, and ventured to offer *conjecturally* five possible causes of the impulse :

1. Sudden fractures of rock, resulting from the steady and slow increase of elevatory pressure.
2. Sudden evolution (under special conditions) of steam.
3. Sudden condensation of steam, also under special conditions.
4. Sudden dislocations in the rocky crust of the earth, through pressure acting in any direction.
5. Occasionally through the recoil due to explosive effects at volcanic foci (p. 79—80).

The first and last of these I am, through subsequent light, disposed now to withdraw or greatly to modify.

The first, the supposed "*snap and jar*, occasioned by
the sudden and violent rupture of solid rock masses,"
to which Mr. Scrope, in his very admirable work on
Volcanoes, is disposed to refer the impulse of earthquake
shocks (Scrope, 2nd edit., p. 294), I believe may be
proved on acknowledged physical principles—when applied to the known elasticities and extensibilities of
rocks, and keeping in view the small thicknesses fractured *at the same instant*—to be capable of only the most
insignificant impulsive effects; and if we also take into
consideration that strata, if so fractured, are necessarily
not *free*, but surrounded by others above and below,
any such impulsive effect emanating from fracture may
be held as non-existent or impossible.   In the statement
of his views which follows, and in objecting to my second
and third possible causes (p. 295—296, headed "Objec-
tions to Mallet's Theory"), Mr. Scrope appears to me
to have fallen into the error of assuming that the nature
of the *impulse*, or the cause producing it, forms any part
of "my theory of earthquake movement," or in anywise
affects it.   I carefully guarded against this in the original
Paper ("Transactions, Royal Irish Academy," Vol. XXI.,
p. 60, and again, p. 97), when I stated "it is quite im-
material to the truth of my theory of earthquake
motion what view be adopted, or what mechanism be
assigned, to account for the original impulse."

As regards the fifth conjecture suggested by me, I
am now, with better knowledge and larger observation
of volcanic phenomena, not prepared to admit any single
explosion at volcanic vents of a magnitude sufficient to
produce by its recoil an earthquake wave of any im-

portance, or extending to any great distance in the earth's crust. The rock of 200 tons weight, said to have been projected nine miles from the crater of Cotopaxi, which I quoted from Humboldt as an example,* I believe to be as purely mythical as the rock (*bloc rejetté*) of perhaps one-sixth of that weight which, previous to the late eruption, lay in the middle of the Atria dell Cavallo, and which it was roundly affirmed had been *blown* out of the crater, but which in reality had at some time rolled down from near the top of the cone, after having been dislodged from some part of the upper lip of the crater walls, where, as its wonderful hardness and texture and its enamel-like surface showed, it had been roasted for years probably.

Nor do I believe in the *sudden* blowing away of one-half the crater and cone of Vesuvius, or of any other volcano, at one effort, however affirmed.

Nothing more than conjecture as to· the nature of the impulse producing great or small Earthquakes can, I believe, as yet be produced. That there is some one master mechanism productive of most of the impulses of great shocks is highly probable, but that more causes than one may produce these impulses, and that the causes operative in small and long repeated shocks, like those of Visp-Comrie and East Haddam, differ

---

* Assuming the point of ejection of this block (the crater) to be 8,000 feet above where it landed, and allowing it as high a density as admissible, and the angle of projection the best for large horizontal range, it may be proved that this mass, to reach nine miles horizontally, would require an initial velocity of projection of from 1,500 to 1,600 feet per second, one as great as that of a smooth-bore cannon-shot at the muzzle, and perfectly inconceivable to be produced by a volcano.

much from those producing great Earthquakes, is almost certain.

We shall be better prepared to assign all of these when we have admitted a true theory of volcanic action, and so are better able to see the intimate relations in mechanism between seismic and volcanic actions.

It is not difficult meanwhile to assign the very probable mechanism of those comparatively petty repercussions which are experienced in close proximity to volcanic vents when in eruption, and which, though certainly seismic in their nature, and powerful enough, as upon the flanks of Etna, to crack and fissure well-built church-towers, can scarcely be termed Earthquakes.

In my First Report I stated that almost nothing was known then of the distribution of recorded Earthquakes in time or in space over our globe's surface, and I proposed the formation and discussion of a complete catalogue of all recorded Earthquakes, with this in view.

This was approved by the Council of the British Association and at once undertaken by me, with the zealous and efficient co-operation of my eldest son, Dr. J. W. Mallet. Nearly the whole of the Second British Association Report, of 1851, is occupied with the account of the experiments as to the transit rate of artificially made shocks in sand and granite, as already referred to.

The Third Report, of 1852—1854, contains the whole of this, "The Earthquake Catalogue of the British Association" (of which, through the liberality of that body, more than one hundred copies were distributed freely), in which are given, in columnar form, the following particulars, from the earliest known dates to the end of 1842:

1. The date and time of day, as nearly as recorded.
2. The locality or place of occurrence.
3. The direction, duration, and number of shocks so far recorded.
4. Phenomena connected with the sea—great sea-waves, tides, etc.
5. Phenomena connected with the land—meteorological phenomena preceding and succeeding. Secondary phenomena—all minor or remarkable phenomena recorded.
6. The authority for the record.

Though most materially assisted by the previous labours and partial catalogues of Von Hoff, Cotte, Hoffman, Merrian, and, above all, of Perrey, the preparation of this catalogue—which demanded visits to the chief libraries of Europe, and the collating of some thousands of authors in various languages and of all time—was a work of great and sustained labour, which, except for my dear son's help, I should never have found time and power to complete. Professor Perrey, formerly of the Faculté des Sciences of Dijon, now *en retrait*, who has devoted a long and useful life to assiduous labours in connection with Seismology, was our great ally; and his catalogues are so large and complete for most known parts of the world after 1842, that we were able to arrest our own catalogue at that date, and take M. Perrey's as their continuation up to 1850.

The whole British Association Catalogue thus em-

braces the long historic period of from 1606 B.C. of
vulgar chronology, when the first known Earthquake
is recorded, to A.D. 1850; and the base of induction
which it presents as to the facts recorded extends to
between 6,000 and 7,000 separate Earthquakes. My
Fourth Report ("Reports, British Association, 1858,")
is occupied principally with the discussion of this great
catalogue, and with that of several special catalogues
produced by other authors with limited areas or objects.

The discussion of M. Perrey's local catalogues with
those of others, in reference to a supposed prevalent
apparent horizontal direction of shock in certain regions
—as to distribution, as to season, months, time of day
or night, relation to state of tide—the bearings of the
views of Zantedeschi and others as to the probable
existence of a terrane tide—the supposed relations of
the occurrence of Earthquakes upon the age of the
moon, as deduced by Perrey, viz.: that 1st, Earthquakes
occur most frequently at the syzygies; 2nd, that their
frequency increases at the perigee and diminishes at
the apogee; 3rd, that they are more frequent when the
moon is on the meridian than when she is 90° away
from it—and the views of several authorities as to the
distribution of Earthquakes in time and in space—
occupy the first 46 pages of this Report.

It then proceeds to discuss the distribution in time
and in space as deduced from the full base of the great
catalogue.

The results as to time are reduced to curves, and
those as to space (or distribution over our globe's
surface) to the great seismic map (Mercator's pro-

jection), upon which and in accordance with certain principles and conventional laws, which admit of the indication of both intensity and frequency, all recorded Earthquakes have been so laid down as to present a real indication of the distribution of seismic energy for the whole historic period and all over the world.

The original of this map, which also shows the Volcano (size, about 7 feet by 5 feet), remains for reference in the custody of the Royal Society. A reduced copy was published with the Report, and to a still more reduced scale has been reproduced in other places. It is impossible here to do more than refer to a few of the more salient points.

As regards distribution in time, durational seismic energy may be considered as probably constant during historic time, though it is probably a decaying energy viewed in reference to much longer periods. It does not appear of the nature of a distinctly periodic force.

1. Whilst the minimum paroxysmal interval may be a year or two, the average interval is from five to ten years of comparative repose.

2. The shorter intervals are in connection with periods of fewer Earthquakes, not always with those of least intensity, but usually so.

3. The alternations of paroxysm and of repose appear to follow no absolute law deducible from these causes.

4. Two marked periods of extreme paroxysm are observable in each century (for the last three centuries), one greater than the other—that of

greatest number and intensity occurring about the middle of each century, and the other towards the end of each.

As respects season, there appear distinct indications of a maximum about the winter solstice, and equally so of a minimum rather before the autumnal equinox. It is not improbable that there is a remote relation between Earthquakes and the annual march of barometric pressure.

We may expect, at present, one great Earthquake about every eight months, and were we possessed of a sufficient report from all parts of our globe, we should probably find scarcely a day pass without a very sensible Earthquake occurring somewhere, whilst, as regards still smaller tremors, it might almost be said that our globe, as a whole, is scarcely ever free from them.

As respects the distribution of seismic energy in space of our earth's surface, it is that of bands of variable and of great breadth, with sensible seismic influence extending to from 5° to 15° transversely, which very generally follow :

1. The lines of elevated tracts which mark and divide the great oceanic or terra-oceanic basins (or *saucers*, as I have called them, from their shallowness in relation to surface, in this discussion) of the earth's surface.

2. And in so far as these are frequently the lines of mountain chains, and these latter those of volcanic vents, so the seismic bands are found to follow these likewise. Isolated Volcanoes are found in these bands also.

3. While sensible seismic influence is generally
   limited to the average width of the band,
   paroxysmal efforts are occasionally propagated
   to great distances transversely beyond that.
4. The sensible width of the band depends upon the
   energy developed at each point of the length,
   and upon the accidental geologic and topographic conditions along the same.
5. Seismic energy *may* become sensible at any point
   of the earth's surface, its efforts being, however,
   greater and more frequent as the great lines of
   elevation and of volcanic activity are approached ;
   yet not in the inverse ratio of distance, for many
   of the most frequently and terribly shaken
   regions of the earth, as the east shore of the
   Adriatic, Syria, Asia Minor, Northern India, etc.,
   are at great distances from active Volcanoes.
6. The surfaces of minimum or of no known disturbance are the central areas of great oceanic
   or of terra-oceanic basins or saucers, and the
   greater islands existing in shallow seas.

Space obliges me to pass unnoticed here many minor
but not unimportant deductions. The discussions as to
distribution in time and space occupy seventy-two pages
of this fourth and last Report, the remainder of which
(thirty-one pages) embraces the description and mathematical discussion as to seismometers, to which I may
refer, as comprising the most complete account of these
instruments that has, I believe, been anywhere given.

The appendix to the Report comprises the entire
bibliography of Earthquakes collected during those

researches, and a concluding chapter on desiderata, and inquiries as to ill-understood phenomena supposed to be connected with Earthquakes.

In 1849-50, I was honoured by the request to draw up the article " Earthquake Phenomena," which has appeared in the first and subsequent editions of the " Admiralty Manual of Scientific Inquiry." Originally the subject was intended to have formed part of the article on Geology, entrusted to Mr. Darwin, who consulted me upon the subject ; and upon my representing how much Earthquakes had, within a short time, become matter for the mathematician and physicist, he, with a singleness of eye to science which it is but just to place on record, took the necessary steps with the Admiralty authorities that Earthquakes should form a separate article, and advised its being placed, as it was, in my hands. To record this will, I believe, be sufficient justification for my reference to this article, in which a good deal of information as to Seismometry is to be found.

By recurring to Mr. Hopkins's Report on Earthquake Theory, before remarked upon (" Report of British Association, 1847 "), it will be seen that the solutions of the problems which he there gives for finding the depth of focus of shock are founded upon the *velocity of propagation* of the wave in the interior of the mass, the *apparent horizontal velocity* and the *horizontal direction of propagation* at any proposed point being known (p. 82).

By this it appears plainly that at that time

Mr. Hopkins supposed that it was the *velocity of translation* of the wave of shock that did the mischief, and not the *velocity of the wave particle*, or wave itself. And, further, that the former might be obtained by reference simply to the modulus of elasticity of the rock of any given formation, as, indeed, was my own earliest view when I produced my " Dynamics of Earthquake " in 1846. From the remarks already made as to the vast difference between the actual transit velocity in more or less discontinuous rocks—such as they occur in Nature—it will be equally obvious that Mr. Hopkins's methods, as above mentioned, are impracticable, even were there no confusion between the velocity of translation of the wave and that of the wave particle or wave itself.

This applies also to the demonstration and diagram (taken from Hopkins) given by Professor Phillips (" Vesuvius," pp. 258—259).

In December, 1857, occurred the great Neapolitan Earthquake, which desolated a large portion of that kingdom ; and an opportunity then arose for practically applying to the problems of finding the directions of earthquake shock at a given point through which it has passed, and ultimately the position and depth of focus, other methods, which I had seen, from soon after the date of publication of my original Paper (1846), were easily practicable, and the details of which I had gradually matured.

Bearing in mind that, in the case of the normal vibration in any elastic solid of indefinite dimensions, the direction of motion in space of the *wave particle* coincides in the first semiphase of the wave, and at the

instant of its *maximum velocity* with the right line
joining the particle and the focus or centre of disturbance, it follows that, in the case of earthquakes, the
normal vibration of the wave of shock is always in a
vertical plane passing through the focus and any
point on the earth's surface through which the
shock passes (assuming for the present no disturbing
causes after the impulse has been given), and that
at such a point the movement of the wave particle
in the first semiphase of the wave is in the same
direction or sense as that of translation; and at the
moment of maximum velocity the direction in space
of the motion of the wave particle is that of the right
line joining the point through which the wave has
passed with the focus or centre of impulse.

If, therefore, we can determine the direction of
motion of the wave particle in the first semiphase, and
its maximum velocity, we can obtain, from any selected
point, a line (that of emergence of the shock) *somewhere in which*, if prolonged beneath the earth, the
focus must have been; and if we can obtain like results for two or more selected points, we decide the
position and the depth of the focus, which must be in
the intersection of the several lines of direction of the
wave particle motion at each point, when prolonged
downwards.

Now, as I have said, it is the *vibration of the wave
itself*, i.e., the motion of the wave particle that does the
mischief—*not* the transit of the wave from place to
place on the surface; just as in the analogous (but *not*
similar) case of a tidal wave of translation running up

an estuary and passing a ship anchored there, it is not
the transit up the channel, but the wave form itself—
i.e., the motion of the wave particles—that lifts the ship,
sends her a little way higher up channel, drops her to her
former level, and sends her down channel again to the
spot she lay in just before the arrival of the wave.

Everything, therefore, that has been permanently
disturbed by an earthquake shock has been thus moved
in the direction and with the maximum velocity impressed upon it by the wave particle in the first semiphase of the wave; and thus almost everything that
has been so disturbed may, by the application of established dynamical principles, be made to give us more
or less information as to the velocity of the wave
particle (or as we, for shortness, say, the velocity of
shock), the direction of its normal vibration, and the
position and depth beneath the earth's surface, from
which came the generating impulse. We thus arrive
at these as simply and as surely as we can infer from
the position taken by a billiard ball, on which certain
forces are known to have acted, the forces themselves
and their direction; or, from a broken beam, the pressure or the blow which fractured it.

It is obvious, then, that nearly every object disturbed,
dislocated, fractured or overthrown by an earthquake
shock is a sort of natural seismometer, and the best
and surest of all seismometers, if we only make a judi-
cious choice of the objects which being found after such
a shock, we shall employ for our purpose. This was the
principle which I proposed to the Royal Society at once
to apply to the effects of the then quite recent great

Neapolitan Earthquake of 1857, and which, through the liberality and aid of that body, I was enabled to employ with the result I had pretty confidently anticipated, namely, the ascertainment of the approximate depth of the focus.

Every shock-disturbed object in an earthquake-shaken country is capable of giving *some* information as to the shock that acted upon it; but it needs a careful choice, and some mechanical νοῦς, to select *proper* and the best objects, so as to avoid the needless perplexity of disturbing forces *not* proper to the shock, or other complications.

When properly chosen, these natural seismometers, or evidences fitted for observation after the shock, are of two great classes, by which the conditions of the earthquake motion are discoverable :

1. Fractures or dislocations (chiefly in the masonry of buildings), which afford two principal sources and sorts of information, namely:

   a. From the observed *directions of fractures or fissures*, by which the *wave path*, and frequently the *angle of emergence*, may be immediately inferred.

   b. Information from the preceding, united with known conditions as to the strength of materials to resist *fracture*, by which the *velocity* of the fracturing impulse may be calculated.

2. The overthrow or the projection, or both, of bodies large or small, simple or complex. From these we are enabled to infer:

   c. By direct observation, the *direction in azimuth* of the wave path.

d. By measurements of the horizontal and vertical distances of overthrow or of projection, to infer either the *velocity* of projection, or *angle of emergence.*

Fractures by shock present their planes always nearly in directions transverse to the wave path. Projections or overthrow take place (unless secondarily disturbed) in the line of the wave path, or in the vertical plane passing throught it : but the direction of fall or over-throw may be either in the same direction as the wave transit (*i. e.,* as the motion of the wave particle in the first semiphase), or contrary to it.

It is thus obvious that the principal phenomena presented by the effects of earthquake shock upon the objects usually occurring upon the surface of the inhabited parts of the earth, resolve themselves into problems of three orders, and are all amenable to mechanical treatment, viz. :

1. Problems relating to the direction and amount of velocity producing fracture or fissures.

2. Problems relating to the single or multiplied oscillations of bodies, considered as compound pendulums.

3. Problems referable to the theory of projectiles.

These three may combine in several cases, and on the part of the observer must combine with measurements, angular and linear, and with geodetic operations to be conducted in the shaken country.

The methods of application in detail are described fully, as well as their actual application and results, in my work published in 1862 (2 vols.), entitled "The

First Principles of Observational Seismology, as developed in the Report to the Royal Society of London of the Expedition made by Command of the Society into the Interior of the Kingdom of Naples, to investigate the Circumstances of the Great Earthquake of December, 1857," to the many illustrations of which the pecuniary grant, in aid, of £300 was most liberally made to the publishers (Messrs. Chapman and Hall) by the Society.

It is not my intention here, nor would space allow, of my going into the details of observation, nor of the deductions and conclusions I have recorded in those volumes. I have referred to their contents as marking the advent of a new method. I have ventured to call it a new *organon* in the investigation of Earthquakes, and, through them, of the deep interior of our earth; and will only add that the method, on this its very first trial, proved fertile and successful. The depth of focus for this shock of December, 1857, was about seven to eight geographical miles below sea level, roughly stated. It gives me great pleasure to add that my friend, Dr. Oldham, Director-General of the Geological Survey of India, has since applied these same methods to the phenomena of the great Cachar Earthquake of the 10th January, 1869, and with success. The pressure of official duties has, he informs me, as yet prevented his fully working out his results, but they appear so far to indicate, as we should expect, a depth of focus or origin considerably greater than in the European case of 1857. Some account of Dr. Oldham's results were this year communicated to the Geological Society of

London through myself, they are of great interest and importance.

Such, briefly and imperfectly sketched, is the existing state of Seismology. As a branch of exact science it is, as it were, an affair of yesterday. It is with reluctance that I have been compelled, in this review, to refer to my own work so prominently. The harvest has been and still is plenteous, but in this field of intellectual work the labourers are few. This must continue to be so as long as Geology shall continue to be viewed in public estimation (in England at least) as a fashionable toy, that everyone who has been to school is supposed capable of handling; and until all who profess to be geologists shall have learnt that, to make sound progress, they must first become mathematicians, physicists and chemists.

It is to the general imperfect knowledge of these sciences amongst geologists that speculative errors show such vitality, and that Geology makes such poor progress towards becoming the interpretation of the world as a machine (*Erdkunde*).

It is for the same reason that Seismology and Vulcanology make little progress; the first cannot be pursued beyond its present boundaries, nor can even its present position be understood or explained by anyone unfamiliar with the laws of wave motion, of all classes of waves; and it would be easy to show, by quoting from various British or foreign text-books on Geology, how extremely imperfect is the grasp of some of the authors upon the subject of earthquake-wave motion, even such as they admit and endeavour to explain and apply: in fact, many geologists appear never to have framed to themselves any clear idea of what *is* a wave of any sort, liquid or elastic. The general silence as to seismic theory of French geological writers is remarkable, to whatever cause attributable. It has been said that French philosophers show themselves little disposed to acknowledge or to follow the lead of their foreign compeers in any branch of science. If this be true, or in so far as it may be so, it is unworthy of French science, which has such boundless claims upon our homage. I am disposed to attribute the fact in this case to other circumstances; and, amongst these, to the small extent to which our language is known amongst French scientific men.

Germany has shown more desire to cultivate this branch of science. Although, as yet, the distinct enunciation of its fundamental principles has but sparsely found its way into her text-books, several able monographs, such as those of Schmidt and of Höttinger, prove how completely some of her philosophers have mastered and how well applied them. The men of science of Northern Italy, amongst whom so many glorious names are to be found on the roll of discovery, have shown themselves quite alive to the importance of Seismology; and I know of no more clear, exact and popular exposition of its principles and application, and of its cosmical relations, than is to be found in a small volume by Professor Gerolamo Boccardo, published at Genoa in 1869, entitled *Sismopirologia Terremoti, Vulcani e lente oscillazione del suolo, saggio di una teoria di Geographia Fisica.*

My object, so far, has been to mark the progress of ascertained theoretic notions as to Seismology. I have, therefore, passed without notice many speculative monographs, and the treatment upon Earthquakes, whether speculative or historical, and however able, that constitutes a prominent feature of nearly all systematic works on Geology.

That which may be at present viewed as achieved and certainly ascertained in theoretic Seismology is the clear conception of the nature of earthquake motion ; the relations to it of great sea or other water wave commotions ; the relations to it of sound waves—as to which, however, more remains to be known ; and the relations of all these to secondary effects, tending in various ways to modify more or less the topographic and other conditions of the land or sea bottom. And in descriptive Seismology the present distribution of the earthquake bands or regions of greatest seismic prevalence and activity are tolerably ascertained, and their connection with volcanic lines and those of elevation rendered more evident. Viewed alone, nothing can yet be said to be absolutely ascertained as to the imme- diately antecedent cause or causes of the impulse. The function of Earthquake, as part of the cosmical machine, has become more clear, as the distinctive boundaries between Earthquake and permanent elevation of the earth have been made evident ; and it has been seen that Earthquake, however contemporaneous occasionally with permanent elevation, is not the cause, though it may be one of the consequences of the same forces which produce elevation ; and thus, that an infinite

number of Earthquakes, however violent, and acting
through however prolonged a time, can never act as an
agent of permanent elevation, unless, indeed, on that
minute scale in which surface elevation may arise from
secondary effects, like that of the Ullah Bund.

Much remains to be done, and much may be expected
even from the continuation, if done in a systematic and
organised manner, of the statistic record of Earthquakes
in connection with those other branches of cosmical
statistics, Climatology, Meteorology, Terrestrial Magnetism, etc., the observation of which is already, to a certain
extent, organised over a large portion of the globe.

And now let us look back for a moment to ask, How,
by what mental path of discovery, have we arrived at
what we have passed in review ?

The facts of Earthquakes have been before men for
unknown ages " open secrets," as Nature's facts have
been well called ; " but eyes had they and saw not."
Facts viewed through the haze of superstition, or of
foregone notions of what Nature *ought* to do, cease to be
facts.   When, after the the great Calabrian Earthquake
of 1783, the Royal Academy of Naples sent forth its
commission of its learned members to examine into the
effects, they had spread around them in sad profusion
all that was necessary to have enabled them to arrive
at a true notion of the nature of the shock, and thence
a sound explanation of the varied and great secondary
effects they witnessed, and of which they have left us
the records in their Report, and the engravings illus-
trative of it.   But we look in vain for any light; the
things seen, often with distortion or exaggeration, are

heaped together as in the phantasmagoria of a wild and terrible dream, from which neither order nor conclusion follow.

Why was this? Why were these eminent *savants* no more successful in explaining what they saw than the ignorant peasants they found in the Calabrian mountains?

Because physical science itself was not sufficiently advanced, no doubt; but also because they had no notion of applying such science as they had, to the very central point itself of the main problem before them, freed from all possible adventitious conditions, and so, as it were, attacking it in the rear. How different might have been the result of their labours, had they begun by asking themselves, What is an earthquake? Can we not try to find out what it *is* by observing and *measuring* what it has done? We see the converse mode of dealing with Nature in Torricelli. " Nature abhors a vacuum," was told him, as the wisdom of his day. Possibly: but her abhorrence is limited, for I find it is *measured* by the pressure of a column of water of thirty-four feet in height. We need not pursue the story with Pascal, up to the top of the Puy de Dôme.

This lesson is instructive generally to all investigators, and particularly here; for Vulcanology, to which we are about now to turn, has occupied until almost to-day much the same position that Seismology did in those of the Neapolitan Commissioners.

Whole libraries have been written with respect to it dealing with *quality*, but *measure* and *quantity* remain to be applied to it.

To a very preponderant class in the civilised world
no knowledge is of much interest or value that does not
point to what is called a " practical result," one measurable into utility or coin.   I do not stop to remark as to
the bad or as to certain good results of this tendency of
mind; but I may venture to point out to all, that the
exact knowledge of the nature of earthquake motion,
even during the short time that it has become known,
has not been barren in results absolutely practical and
utilitarian.   The minute investigation of the destruction
of buildings, etc., and the deductions that have been
made as to the relations between the form, height, materials, methods of building, combination of timber and
of masonry, and many other architectural or constructive
conditions, have made it certain now that earthquake-
proof houses and other edifices can be constructed with
facility, and at no great increase, if any at all, of cost.   I
can affirm that there is no physical necessity why in frequently and violently shaken countries, such as Southern
Italy or the Oriental end generally of the Mediterranean,
victims should hereafter continue by thousands to be
sacrificed by the fall of their ill-designed and badly
built houses.

Were a " Building Act " properly framed, put in
force by the Italian Government in the Basilicatas and
Capitanata, etc., so that new houses or existing ones,
when rebuilt, should be so in accordance with certain
simple rules, a not very distant time can be foreseen
when Earthquakes, passing through these rich and
fertile but now frequently sorely afflicted regions, should
come and go, having left but little trace of ruin or

death behind. Some disasters there must always be, for we cannot make the flanks of mountains, nor the beds of torrents, etc., always secure; but the main mortality of all Earthquakes is in the houses or other inhabited buildings. Make these proof, and the wholesale slaughter is at an end.

The principles we have established have been thus practically applied in another direction. The Japanese Government, with the keen and rapid perception of the powers inherent in European science which characterises now that wonderful people, has commenced to illuminate its coasts by lighthouses constructed after the best European models. But Japan is greatly convulsed by earthquakes, and lighthouses, as being lofty buildings, are peculiarly liable to be destroyed by them.

The engineer of the Japanese Government for these lights, Mr. Thomas Stevenson, C.E. (one of the engineers to the Commissioners of Northern Lights), was instructed to have regard, in the design of those lighthouses, to their exposure to shock. I was consulted by Mr. Stevenson as to the general principles to be observed; and those edifices have been constructed so that they are presumedly proof against the most violent shocks likely to visit Japan; not, perhaps, upon the best possible plan, but upon such as is truly based upon the principles I have developed. Mr. Stevenson has published some account of their construction.

The earthquake regions of South America might with incalculable benefit apply those ideas; and, indeed, they have been, to some extent, already applied by my friend, Mr. William Lloyd, Member of the

Institution of Civil Engineers, to the New Custom Houses constructed from his designs at Valparaiso.

As one of these utilitarian views, and an important one, it will occur to many to ask—Can the moment of the occurrence or the degree of intensity of earthquake shock be predicted, or is it probable that at a future day we may be able to predict them? At present, any prediction, either of the one or the other, is impossible; and those few who have professed themselves in possession of sufficient grounds for such prediction are deceivers or deceived. Nor is it likely that, for very many years to come, if ever, science shall have advanced so as to render any such prediction possible; but it is neither impossible nor improbable that the time shall arrive when, within certain, perhaps wide, limits as to space, previous time, and instant of occurrence, such forewarnings may be obtainable.

Earthquakes, like storms and tempests, and nearly all changes of weather, are not periodic phenomena, nor yet absolutely uncertain or, so to say, accidental as to recurrence.

They are quasi-periodic, that is to say, some of their conditions as to causation rest upon a really periodic basis, as, for example, the recurrence of storms upon the periodic march of the earth, and sun and moon, etc., and the recurrence of Earthquakes upon the secular cooling of our earth; but the conditions in both are so numerous and complicated with particulars, that we cannot fully analyse them—hence, cannot reduce the phenomena to law, and so cannot predict recurrence.

Yet storms and tempests—which were, along with pestilences and Earthquakes, amongst the natural phenomena which Bishop Butler deemed in his own day impossible of human prediction—have already, through the persistent and systematised efforts of meteorological observers, become to a certain extent foreseeable; and medical science assures us that it has rendered that, though to a much less degree of probability, true of pestilences.

We may, therefore, give the utilitarian some hope, that if he will help us along—who value our accessions of knowledge primarily upon a different standard to his —in our task of discovery, our posterity, in a century or two hence, may not improbably possess the advantage of being able, in some degree, to predict their Earthquakes. I fear the inducement will go but a small way with the utilitarian generation, whose bent tends much towards asking, " What has posterity ever done for them ?"

But though we cannot as yet predict the time when an Earthquake may take place in any locality, we can, on mixed statistic and dynamic grounds, in many cases state the limits of probable violence of the next that may recur. For example, the three shafts of marble columns of the Temple of Serapis, at Pozzuoli, each of about $41\frac{1}{2}$ feet in height, and 4 feet 10 inches in diameter at the base, remain standing alone, since they were uncovered, in the year 1750.

Now, as we can calculate exactly what velocity of earthquake-wave motion would be required to overset these, we are certain that, during the last one hundred and twenty-two years, the site of the Temple, and we may say Naples and the Phlegræan fields generally, have never experienced a shock as great as the very moderate one that would overset these columns. A shock whose wave particle had a horizontal velocity of only about 3½ feet (British) per second would overturn these columns; which is only about one-fourth the velocity (within the meizoseismic area) of the great shock of 1857, that produced wide-spread destruction in the Basilicatas, and not enough to throw down any reasonably well-built house of moderate height.

Naples, so far as Earthquake is concerned, whether coming from the throes of Vesuvius or elsewhere, has a pretty good chance of safety. She may possibly (though not probably) be some day smothered in ashes; but is in little danger of being shaken to the earth. During this time there have been taking place, larger eruptions of Vesuvius and earthquake shocks from other centres, together probably about the same number of times as the numbers of those years, when those columns have been more or less shaken.

We may therefore affirm that the probability (on the basis of this experience *only*) is, say 120 to 1, that the next shock, whether derived from Vesuvius, or else-where, that may shake Pozzuoli, will be one less in power than would be needed to overturn the shafts of the Temple of Serapis there.

Let us now turn to the second branch of our subject—viz., Vulcanology—upon which, as yet, we have secured less firm standing ground than we have seen we possess in Seismology, for which reason we took that first into consideration.

It is the part of Vulcanology to co-ordinate and explain all the phenomena of past or present times visible on our globe which are evidences of the existence and action, whether local or general, of temperatures within our globe greatly in excess of those of the surface, and which reach the fusing points of various mineral compounds as found arriving, heated or fused, at the surface.

The stratigraphic geologist sees that such heated or fused masses have come up from beneath, throughout every epoch that he can trace; but he cannot fail to discern more or less a change in the order or character of those outcomings, as he traces them from the lowest and oldest formations to those of the present day. He sees immense outpourings of granitoid or porphyrytic rocks that have welled up and overflowed the oldest strata—huge dykes filling miles of fissures that had been previously opened for the reception of the molten matter that has filled them, and often passing through those masses of previously outpoured rock; later he sees huge tables of basaltic rock poured forth over all. One grand characteristic common to all these—commonly called plutonic products—being that, whether they were poured forth over the surface or injected into cavities in other rocks, the movements of the fused material were, on the whole, hydrostatic and *not explosive.*

At the present day, whatever other evidences we have of high temperature below our globe's surface, that which primarily fixes the eye of the geologist is the

Volcano, whose characteristic, as we see it in activity, *is explosive*. But though there is this great characteristic difference between the plutonic and the volcanic actions and their products, the two, when looked at largely, are seen so to inosculate, that it is impossible not to refer them to an agency common to both, however changed the modes of its action have been between the earliest epochs of which traces are presented to us and the present day.

To us little men, who, as Herschell has well said, in referring to the methods of measuring the size of our globe, " can never see it all at once, but must creep like mites about its surface," the Volcano, in the stupendous grandeur of its effects, tends to fix itself in our minds in exaggerated proportions to its true place in the cosmic machine; and, in fact, nearly all who have sought to expound its nature and mode of origination have occupied themselves far too exclusively with describing and theorising upon the strange and varied phenomena which the volcanic cone itself and its eruptions present, and too often, in the splendour and variety of these, have very much lost sight of what ought to be the centre-point of all such studies, namely, to arrive at some sound knowledge of what is the *primum mobile* of all these wonderful efforts. Nor has the distinction been very clearly seen between the main phenomena presented at and about volcanic active mouths, which can be employed to elucidate the nature of the causation at work far below, and those most varied and curious, and in other respects most pregnant and instructive phenomena, mechanical and chemical,

which are called into action in and by the ejected
matter of the volcanic cone after its ejection. It can
help us but little or very indirectly, in getting at a true
conception of the nature and source of the heat itself
of the Volcano, to examine, for example, all the curious
circumstances that are seen in the movements and
changes in the lava that has already flowed from its
mouth; but it would be of great importance if we can
ascertain, by any form of observation around the cone,
from what depth it has come, or at what depth the
igneous origin lies.

The physician, endeavouring to ascertain the real nature of small-pox or measles, will scarcely make much
progress who, however curiously or minutely, confines
his attention to the pustules that he sees upon the skin.

Yet the Volcano, or rather all volcanic activity as
now operative upon our globe, is, as it were, an experiment of Nature's own perpetually going on before us,
the results of which, if well chosen—that is, as Bacon
says, by keeping to the main and neglecting the acci-
dents—can, when colligated and correctly reasoned upon,
in relation to our planet as a whole, give us the key to
the enigma of terrestrial Vulcanicity in its most general
sense, and at every epoch of our world's geognostic
history, and show us its true place and use in the cosmical machine. Let us glance at the history of past
speculation on this subject, from which so little real
knowledge is to be derived, and then at the salient
facts of Vulcanology as now seen upon our earth, and
finally see if we can connect these with other great cos-

E

mical conditions, so as to arrive at a consistent explanation in harmony with all.

We gain nothing absolutely from the knowledge of the so-called "ancients" as to Volcanoes in Europe at least, where alone historic records likely to refer to them exist. The Volcanoes of Europe are few and widely scattered. The Greeks saw but little of them, and the Romans were all and at all times most singularly unobservant of natural phenomena.

Cæsar never mentions the existence in France of the Volcanoes of Auvergne, so much like those he must have seen in Italy and Sicily; and Roman writers pass in silence that great volcanic region, though inhabited by them, and their language impressed upon the places, as Volvic (*volcano-vicus*) seems with others to indicate; and though there is some reason to believe that one or other of the Puys was in activity within the first five hundred years of our epoch, the notices which Humboldt and others have collected as from Plato, Pausanius, Pliny, Ovid, etc., teach nothing.

Whatever of mere speculation there may have been, volcanic theory, or what has passed for such, there was none before 1700, when Lémery brought forward a trivial experiment, the acceptance of which, even for a moment, as a sufficient cause for volcanic heat (and it retarded other or truer views for years), we can now only wonder at. Breislak's origin, in the burning of subterranean petroleum or like combustibles, was scarcely less absurd than Lémery's sulphur and iron filings.

Davy, in the plenitude of his fame, and full of the

intense chemical activities of the metals of the alkalies which he had just isolated, threw a new but transient verisimilitude upon the so-called chemical theory of Volcanoes, by ascribing the source of heat to the oxidation of those metals assumed to exist in vast, unproved and unindicated masses in the interior of the earth. But Davy had too clear an intellect not to see the baseless nature of his own hypothesis, which in his last work, the " Consolations in Travel," he formally recanted ; and it only survived him in the long-continued though unconvincing advocacy of Dr. Daubeny. So far, the origin of the heat had been sought always, in the crude notion of some sort of *fuel consumed*, whether that were petroleum or potassium and sodium ; but as no fuel was to be found, nor any indicated by the products, so far as known, of the volcanic heat, so what has been called the mechanical theory, in a variety of shapes, took its place.

This, in whatever form, takes its lava and other heated products of the volcano ready made from a universal ocean of liquid material, which it supposes constitutes the interior or nucleus of our globe, and which is only skinned over by a thin, solid crust of cooled and consolidated rock, which was variably estimated at from fourteen to perhaps fifty miles in thickness. Here was a boundless supply of more than heat, of hot lava ready made, the existence of which at these moderate depths the then state of knowledge of hypogeal temperature, which was supposed to go on increasing with depth at the rate of about 1° Fahrenheit, for every thirty or forty feet, seemed quite to sustain.

The difficulty remained, how was this fiery ocean brought to the surface or far above it? To account for this two main notions prevailed, and, indeed, have not ceased to prevail. Some unknown elastic gases or vapour forced it up through fissures or rents pre-existent, or produced by the tension of the elastic and liquid pressure below.

The form in which this view took most consistency, and approaching most nearly to truth, finds the elastic vapour in steam generated from water passed down through fissures from the sea or from the land surface. But to this the difficulty was started, that fissures that could let down water would pass up steam. The objection, when all the conditions are adequately considered, has really no weight; and it has been completely disposed of, since within a few years it has been proved that capillary infiltration goes on in all porous rocks to enormous depths, and that the capillary passages in such media, though giving free vent to water —and the more as the water is warmer—are, when once filled with liquid, proof against the return through them of gases or vapours. So that the deeply seated walls of the ducts leading to the crater, if of such material, may be red hot and yet continue to pass water from every pore (like the walls of a well in chalk), which is flushed off into steam that cannot return by the way the water came down, and must reach the surface again, if at all, by the duct and crater, overcoming in its way whatever obstructions they may be filled with.

And this remarkable property of capillarity sufficiently shows how the lava—fused below or even at or above the level of infiltration—may become interpenetrated throughout its mass by steam bubbles, as it usually but not invariably is found to be.

Nor is it difficult to see such a mechanism between volcanic ducts and fissures conveying down water, as large and open pipes, for a large part of their depth, as shall bring down water to foci of volcanic heat, without the power of the water flowing back except as steam and through the crater.

Indeed, the facts known as to geysers, and those of half-drowned-out Volcanoes such as Stromboli—whose action is intermittent just as much as that of a geyser—show that this is not merely probable. There is, therefore, no need for the hypothesis of those who have supposed all the huge volumes of steam blown off from Volcanoes in eruption to come from vesicular water pre-existent in the minute cavities of crystalline or other rocks before their fusion into lava : a fact not proved for many classes of rock, and for none in sufficient quantity to account for the vast volume of steam required and for the irregularity of its issue.

It is rather to anticipate, but I may state at once that, so far as the admission of superficial waters to the interior, and to any depth to which fissures or dislocation can extend, I believe no valid physical or mechanical difficulties exist, taking into account all the conditions that may come into play together.

Another set of views has been suggested and supported by various writers, which proposes to account for the rise of lava on purely hydrostatic principles. The solid crust, fractured into isolated fragments by tensions due to its own contraction, is supposed to sink into the sea of lava on which it floats; and much ingenuity has been expended in imagining the mechanism by which, in places, the liquid matter is supposed to rise *above* the surface of the crust.

I have no space for discussing these views further than to assert that, in the existing state of our globe, and even admitting a solid crust of only 60,000 metres thick, dislocation of the crust by *tension* is not possible. The solid crust of our globe, as I hope we shall see further on, is not in a state of tension, and has not been so since it was extremely thin, a mere pellicle as compared with the liquid nucleus, but is, on the contrary, in a state of *tangential compression*.

However tenable, in other respects, may be the volcanic theory which rests upon the assumption of a very *thin* crust and a universal ocean of fused rock beneath, it fails wholly to explain many of the most important circumstances observable as to the distribution and movements of existing Volcanoes on our globe.

It affords no adequate explanation of the configuration of the lines of Volcanoes, nor of their occurrence in the ocean bed, nor of their existence in high latitudes, near the Poles, where, no matter how or at what rate our globe cooled from liquidity, the crust must be thickest; nor of the independence of eruptive action of closely adjacent volcanic vents; nor of the non-periodicity, the sudden awakening-up to activity, the as sudden exhaustion, the long repose, the gradual decay of action at particular vents, and of much more that might be stated and sustained as difficulties left by that theory unexplained, or that are of a nature even opposed to it.

The researches of the last few years have, however, as it appears to me, rendered any theory that demands as its postulates a *very thin crust*, and a universal liquid nucleus beneath it, absolutely untenable.

Without attaching any importance to the arguments of Mr. Hopkins, based upon precession and nutation, it appears to me, on various other grounds, some of which have been urged by Sir William Thompson, that the earth's solid crust is not a thin one, at least not thin enough to render it conceivable that water can ever gain admission to a fluid nucleus, if any such still exist, situated at so great a depth; and without such access we can have no Volcano. It is not necessary to go to the extent of a crust of 800 or 1,000 miles thick: with one of half the minor thickness, I believe it may be proved, on various grounds, hydraulic amongst others, that neither water could reach the nucleus, nor the liquid matter of the nucleus reach the surface. Mr. Hopkins having proved to his own satisfaction an enormous thickness for the crust, and seeing clearly the difficulties that this involved to the generally accepted volcanic theory, and having no other to substitute for it, fell back upon that most vague and weak notion of the existence of isolated lakes of liquid rock, existing at comparatively small depths beneath the earth's surface within the solid and relatively cold crust, each supplying its own Volcano, or more than one, with ready-made lava.

What is to produce these lakes of fused matter in the
midst of similar solidified matter? what is perpetually to
maintain their fluidity in the midst of solid matter continually cooling? what has given them their local position? why near or less near the surface? what should
have arranged them in directions stretching in some
cases nearly from Pole to Pole?

Surely this creation of imaginary lakes, merely because it happens to fit the vacant chink that seems
needed to wedge up a falling theory, is an instance of
that abuse of hypothesis against which Newton so vehe-
mently declaims—" *Hypotheses non fingo.*"

Hypothesis, to be a philosophic scaffolding to knowledge, must, as Whewell has said, " be close to the facts,
and not merely connected with them by arbitrary and
untried facts." Yet this appears accepted by Lyell
(10th edition, Vol. II., p. 227, and elsewhere); by
Phillips (" Vesuvius," pp. 331, 332); by Scrope, if, as
I hope, I mistake him not (" Volcanoes," pp. 265,
307—8); though none of these excellent authorities
seem either quite clear or quite satisfied with the notion;
and in the very passage referred to, Lyell *may* have
possibly a much more philosophic notion in view, where
he says: " It is only necessary, in order to explain the
action of Volcanoes, to *discover some cause which is
capable of bringing about such a concentration of heat as
may melt one after the other certain portions of the solid
crust,* so as to form seas, lakes or oceans of subterraneous lava." (Vol. II , pp. 226, 227). If by this is
meant, that all that is needed to complete a true theory
of volcanic action is to discover *an adequate cosmical*

cause for the heat—that is to say, a prime mover to which all its phenomena may be traced back, which shall be at once reconcilable with the conditions of our planet as a cooling mass in space and with facts of Vulcanology as they are now seen upon it—then I entirely agree with it.

It has been my own object to endeavour to discover and develope that adequate cause in a Paper " On Volcanic Energy, an Attempt to develope its True Nature and Cosmical Relations," read (in abstract) before the Royal Society of London (" Proceedings, Royal Society," Vol. XX., May, 1872), and now (October, 1872) under consideration of Council with a view to publication.

I propose concluding this review of the progress of Vulcanology (in which I have had to limit myself to reviewing merely the chief stages of advance towards knowledge of the nature and origin of volcanic heat itself, and have had to pass without notice the vast and important mass of facts and reasonings collected by so many labourers as to its visible phenomena and products, and the still greater mass of speculation, good and bad, on every branch of the subject), by giving a necessarily very brief and imperfect sketch of my own views as in that Paper in part developed. It will first be necessary to retrace our steps a little, in order to gain such a point as shall afford us a fuller view of the whole problem before us.

It is not necessary to dilate, even did space allow, upon the many points which bind together Earthquakes and Volcanoes as belonging to the play of like forces. These are generally admitted; and in various ways, more

or less obscure, geologists generally have supposed some
relations between these and the forces of elevation,
which have raised up mountain chains, etc.

No one, however, that I am aware of, prior to myself,
in the Paper just alluded to, has attempted to show,
still less to prove upon an experimental basis, that all
the phenomena of elevation, of volcanic action, and
of Earthquakes, are explicable as parts of one simple
machinery—namely, the play of forces resulting from
the secular cooling of our globe. We have seen that,
on the whole, both Earthquakes and Volcanoes follow
along the great lines of elevation of our surface. Any
true solution of the play of forces which has produced
any one of those three classes of phenomena must
connect itself with them all, and be adequate to account
for all. And this would have earlier been seen, had
geologists generally framed for themselves any correct
notions of the mechanism of elevation itself, and seen
its real relation with the secular cooling of our planet.
But the play of forces resulting from this secular cooling
has never, until very recently, been adequately or truly
stated. The arbitrary assumption and neglect of several
essential conditions by La Place, in his celebrated Paper
" On the Cooling of the Earth," in the fifth volume of the
"Mécanique Céleste," and the arbitrary and unsustainable
hypothesis of Poisson upon the same subject, have tended
to retard the progress of physical Geology as to the nature
of elevation : the first, by leaving the geologist in doubt as
to whether our globe were cooling at all ; the second, by
suggesting distorted notions as to the mode of its cooling
and consolidation. On the other hand, neither geologists

nor mathematicians generally have framed for themselves any clear notions of the mechanism of elevation.
Had a true conception been formed of the forces and
interior movements brought necessarily into operation
by the secular cooling of the globe, geologists could
scarcely have failed to see that their notion as to the
way and direction in which the forces producing elevation have actually acted could not, if arising from
refrigeration, be those which they have almost univer-
sally supposed, namely—some force acting vertically upwards, *i. e.*, radially from the centre of the sphere. Had
geologists only looked at Nature with open eye, they
must have seen that mountain ranges, and elevations
generally (exclusive of volcanic cones), presented circumstances absolutely incompatible with their having
been thrust up by any force *primarily* acting in the
direction of a radius to the spheroid.

Yet this is the erroneous notion of the mechanism of
elevation which to the present hour prevails amongst
geologists, so far as they in general have framed to
themselves any distinct idea of such mechanism at all.

Thus, only to cite two examples from recent authors
of justly high reputation. Lyell says of the probable subterranean sources, whether of upward or down-
ward movement, when permanently uplifting a country,
and in reference to the crumpling of strata on mountain flanks by lateral pressure, it would be rash to
assume these able to resist a power of such stupendous
energy, "*if its direction, instead of being vertical*, happened
to be oblique or horizontal." This is somewhat vague—
and I trust I do not mistake or misrepresent the illustrious

author—yet it is the most explicit expression I can find
in the " Principles of Geology " as to his notion of the
primary direction of elevatory force (Edit. 10, Vol. I.,
p. 133). That Mr. Scrope's idea is that only of primary
radial or vertical direction of such forces, is apparent on
inspecting his Diagram No. 64 (" Volcanoes," p. 285),
and in the use of the words, "an axial wedge of
granite," which, on the next page, we find is "liquefied
granite ; " and if we read on to page 294, and refer also
to pages 50 and 51, I believe there can be no doubt that
vertical or *direct up-thrust* is the author's notion of the
primary direction of all forces of elevation. The true
nature of these forces was, however, clearly seen and
most justly stated by Constant Prevost (" Compt.
Rend.," Tome XXXI., 1850, and " Bulletin de la Société
Géolog. de France," Tome II., 1840) as consisting, not
in forces of some unknown origin acting primarily in
the vertical, but in *tangential pressures acting horizontally, and resolved by mutual pressures at certain
points into vertical resultants.* These Prevost rightly
attributed to the contraction of the earth's solid crust.
The same idea has been adopted by Elie de Beaumont
as the true mechanism of the elevation of mountain
ranges ; and although De Beaumont's views as to the
thinness he assigns to the solid and contracting crust,
and his strange deduction as to the parallelism of contemporaneous mountain chains uplifted by its spas-
modic action along certain lines, may be untenable, his
notion generally as to the play of forces producing
mountain elevation is much more nearly correct.

Mr. Hopkins's notion is simply that of the geologists.

Anyone who reads his well-known papers on elevation and the formation of fissures, etc., must see that he views all elevatory forces as of liquids or quasi-liquids forced up and acting primarily *vertically* upon the strata above them, and that these strata are not under tangential compression, but under tension. Hence the mathematical deductions contained in those papers as to the directions in which elevatory forces act, and in which fissures are formed by them, are not in any way a setting forth of such facts as occur in Nature, and, much attention as they have attracted, can only now be viewed as exercises of mathematical skill misapplied, because based upon data not to be found in Nature. In fact, those papers do but misrepresent Nature, and, like many other mathematical investigations based on untrue or insufficient data, have tended to retard knowledge.

The views which I have put forward in the Paper I have referred to, read to the Royal Society, recapitulated in skeleton, so to say, are as follows. Omitting those portions which treat of our globe from the period of the first liquefaction out of a nebulous condition, and of the earliest stages of the cooling by radiation into space, when the crust was extremely thin, and of the deformation of the spheroid as one of the first effects of its contraction, and through that the general shaping out of continents and ocean beds; I have endeavoured to show that the rate of contraction of the crust, while very thin, exceeded that of the large fluid nucleus supporting it, and so gave rise to *tangential tensions* in the crust, and fracturing it into segments; next, that as the crust thickened, these *tensions* were

gradually converted into *tangential pressures*, the contraction of the nucleus now beginning to exceed (for
equal losses of heat) that of the crust through which
it cooled. At this stage these tangential pressures gave
rise to the *chief* elevations of mountain chains—not
by liquid matter by any process being injected from
beneath vertically, but by such pressures, mutually
reacting along certain lines, being resolved into the
vertical, and forcing upwards more or less of the crust
itself. The great outlines of the mountain ranges and
the greater elevation of the land were designated and
formed during the long periods that elapsed in
which the continually increasing thickness of crust
remained such that it was still, as a whole, flexible
enough, or opposed sufficiently little resistance to crushing,
to admit of this uprise of mountain chains by resolved
tangential pressures.   I have shown that the simple
mechanism of such tangential pressures is competent to
account for all the complex phenomena both of the
elevations and of the *depressions* that we now see on the
earth's surface (other than continents and ocean beds),
including the production of gaping fissures (in directions
generally orthogonal to those of tangential pressure).
And as our earth is still a cooling body, and the crust,
however now thicker and more rigid, is still incapable of
sustaining the tangential pressures to which it is now
exposed, so I by no means infer that slow and small
(relatively) movements of elevation and depression may
not be still and now going on upon the earth's surface ;
in fact all the phenomena of elevation and depression,
rending, etc., which at a much remoter epoch acted

upon a much grander and more effective scale. So that, for aught my views say to the contrary, all the mountain chains in the world may be possibly increasing in stature year by year, or at times; but in any case at a rate almost infinitesimally small in its totality over the whole earth to that with which their ridges were originally upreared.

But the thickness of the earth's crust—thus constantly added to, by accretion of solidifying matter from the still liquid or pasty nucleus, as the whole mass has cooled—has now assumed such a thickness as to be able to offer a too considerable resistance to the tangential pressures, to admit of its giving way to any large extent by resolution upwards; yet the cooling of the whole mass is going on, and contraction, though unequal, both of thick crust and of hotter nucleus beneath also, whether the latter be *now* liquid or not. Were the contraction, lineal or cubical, for equal decrements or losses of heat, or in equal times—equal both in the material of the solidified crust and in that of the hotter nucleus—there could be no such tangential pressures as are here referred to, at any epoch of the earth's cooling. But in accordance with the facts of experimental physics, we know that the co-efficient of contraction for all bodies is greater as their actual temperature is higher, and this both in their solid and liquid states.

Hence for equal decrements of heat, or by the cooling in equal times, the hotter nucleus contracts more than does its envelope of solid matter.

The result is now, as at all periods since the signs changed of the tangential forces thus brought into play— *i.e.*, since they became tangential *pressures*—that the nucleus tends to shrink away as it were from beneath the crust, and to leave the latter, unsupported or but partially supported, as a spheroidal dome above it.

Now what happens? If the hollow spheroidal shell were strong enough to sustain, as a spheric dome, the tangential thrust of its own weight and the attraction of the nucleus, the shell would be left behind altogether by the nucleus, and the latter might be conceived as an independent globe revolving, centrally or excentrically, within a shell outside of it. This, however, is not what happens.

The question then arises, Can the solid shell support the tangential thrust to which it would be thus exposed? By the application to this problem of an elegant theorem of Lagrange, I have proved that it cannot possibly do so, no matter what may be its thickness nor what its material, even were we to assume the latter not merely of the hardest and most resistant rocks we know anything of, but even were it of tempered cast-steel, the most resistant substance (unless possibly iridio-osmium exceed it) that we know anything about. Lagrange has shown that if P be the normal pressure upon any flexible plate curved in both directions, the radii of these principal curvatures being ϱ' and ϱ'', and T the tangential thrust at the point of application and due to the force P, then:

$$P = T\left(\frac{1}{\varrho'} + \frac{1}{\varrho''}\right)$$

When the surface is spherical, or may be viewed as such, $\varrho' = \varrho''$ and

$$P = \frac{2\,T}{\varrho} \text{ or, } T = P \times \frac{\varrho}{2}$$

In the present case P is for a unit square (taken relatively small and so assumed as plane) of the shell, suppose a square mile, equal to the effect of gravity upon that unit, ϱ being the earth's radius, and if we assume the unit square be also a unit in thickness, P is then the weight of a cubic mile of its material; and if we take (roughly) the earth's radius as 4,000 miles, the tangential pressure, T, is, on *each face* of the cubic mile, equal to

$$\frac{4000}{2}\ \mathrm{P},$$

or equal to the pressure of a column of the same material of 2,000 times its weight.

If the cubic mile that we have thus supposed cut out of the earth's crust at the surface were of the hardest known granite or porphyry, it would be exposed to a crushing tangential pressure equal to between 400 and 500 times what it could withstand, and so must crush, even though only left unsupported by the nucleus beneath, to the extent of $\frac{1}{400}$ or $\frac{1}{500}$ of its entire weight. And what is true here of a mile taken at the surface, is true (neglecting some minute corrections for difference in the co-efficient of gravity, etc.) if taken at any other depth within the thick crust.*

---

* The Rev. O. Fisher, M.A., F.G.S., in a most interesting and valuable Paper, "On the Elevation of Mountain Chains by Lateral Pressure, its Cause, and the Amount of it, with a Speculation on the Origin of Volcanic Action," read, April, 1868, and published in the Transactions of the Cambridge Philosophical Society, Vol. XI., Part III., in 1869, has deduced the necessary crushing of the earth's crust by a different but closely analogous method. I had not seen this Paper until after my own was in the hands of the Royal Society. The author's volcanic views are wholly different from my own, and do not appear to me equally valid with his notions as to elevation.—R. M.

The crust of our earth, then, as it now is, must crush, to follow down after the shrinking nucleus—if so be that the globe be still cooling, and constituted as it is; even to the limited extent to which we know anything of its nature—it must crush unequally, both regarded superficially and as to depth; generally the crushing lines being confined to the planes or places of greatest weakness; and the crushing will not be absolutely constant and uniform anywhere, or at any time, or at any of those places of weakness to which it will be principally confined, but will be more or less irregular, quasi-periodic, or paroxysmal: as is, indeed, the way in which all known material substances (more or less rigid) give way to a slow and constantly increasing, steady pressure.

We have now to ask, *How much* of this crushing is going on at present year by year? And the answer to this depends upon what amount of heat our world is losing into space year by year.

Geologists who have taken on trust the statement, that La Place has proved that the world has lost no sensible amount of heat for the last 10,000 years seem generally to suppose that to be a fact; but in reality La Place has *proved* nothing of the sort, as those geological teachers who have echoed the conclusion should have known, had they deciphered the mathematical argument upon which it has been supposed to rest.

By application of Fourier's theorem (or definition) to the observed rate of increment of heat in descending from the geothermal *couche* of invariable temperature, and the co-efficients of conductivity of the rocks of our earth's crust, as given by the long-continued observations made beneath the Observatories of Paris and of Edinburgh, it results that the annual loss of heat into space of our globe at present is equal to that which would liquefy into water, at 32° Fahr., about 777 cubic miles of ice; and this is the measuring unit for the amount of contraction of our globe now going on. The figures are not probably exact, for the data are not on a basis sufficiently full or exactly established as yet; but they are not very widely wrong, and their precise exactness is not material here. Now, how is this annual loss of heat (great or small, as we may please to view it) from the interior of our globe disposed of?

What does it *do* in the interior? We have already seen that it is primarily disposed of by conversion into work; into the work of diminishing the earth's volume as a whole, and in so doing crushing portions of the solid surrounding shell.

But does the transformation of lost heat into the work of vertical descent, and of the crush as it follows down after the shrinking nucleus, end the cycle? No. A very large portion of the mechanical work thus produced, and resolved, as we have seen, into tangential crushing pressure, is retransformed into heat again in the very act of crushing the solid material of the shell. If we see a cartload of granite paving-stones shot out in the dark, we see fire and light produced by their collision; if we rub two pieces of quartz together, and crush thus their surfaces against each other, we find we heat the pieces and evolve light.

The machinery used for crushing by steam-power,

hard rocks into road metal, gets so hot that the surfaces cannot be touched.

These are familiar instances of one result of what is now taking place by the crushing of the rocky masses of our cooling and descending earth's crust, every hour beneath our feet, only upon a vastly greater scale. It is in this local transformation of work into heat that I find the true origin of volcanic heat within our globe. But if we are to test this, so as in the only way possible to decide is it a true solution of this great problem, we must again ask the question, *How much?* and to answer this, we must determine *experimentally* how much heat can be developed by the crushing of a given volume, say a cubic mile, of such rocky materials as we know must constitute the crust of our globe down to the bottom of the known sedimentary strata, and extending to such crystalloid rocks as we may presume underlie these. We must also obtain at least approximately what are the co-efficients of *total contraction* between fusion and atmospheric temperature of such melted rocks, basic and acid silicates, as may be deemed representative of that co-efficient for the range of volcanic fused products, basalts, trachytes, etc., which probably sufficiently nearly coincide with that of the whole non-metallic mass of our globe.

The first I have determined experimentally by two different methods, but principally by the direct one of the *work* expended in crushing prisms of sixteen representative classes of rock; the specific gravities and specific heats of which I have also determined.

If H be the height of a prism of rock crushed to powder by a pressure, P, applied to two opposite faces,

which, when the prism has been reduced to its volume
in powder, has acted through a range of $H - t$, then

$$\frac{P \times (H - t)}{772}$$

is the heat corresponding to the work expended in the
crushing, expressed in British units of heat. The following were the rocks experimented upon: Caen stone, Port-
land (both oolites), magnesian limestone, sandstones of various sorts, carboniferous limestones (marbles), the older
slates (Cambrian and Silurian),basalts, various granites and
porphyries, thus ranging from the newest and least resistant to the oldest and most resistant rocks. The results
have been tabulated, and are given in detail in my Paper,
now in possession of the Royal Society. The minimum
obtained is 331 and the maximum 7,867 British units
of heat developed, by transformation of the work of
crushing one cubic foot of rock. If we apply the
results to a thickness of solid crust of 100 miles (British),
of which the upper twenty-one miles consist of neozoic,
newer palæozoic, older palæozoic and azoic rocks in
nearly equal proportion as to thickness, and the
remaining eighty miles of crystalloid rocks (acid and
basic magmas of Durocher) of physical properties
which we may assume not very different from those
of our known granites and porphyries—and which, in
so far as they may differ, would give a still *higher*
co-efficient of work transformed into heat than I have
attributed to them by ranging them as only equal to the
granites, etc.—then we obtain a mean co-efficient for the
entire thickness of crust of 100 miles of 6,472 British
units of heat, developable from each cubic foot of
its material, if crushed to powder. It results from this

that each cubic mile of the mean material of such a crust, when crushed to powder, developes sufficient heat to melt 0·876 cubic miles of ice into water at 32°, or to raise 7·600 cubic miles of water from 32° to 212° Fahr., or to boil off 1·124 cubic miles of water at 32° into steam of one atmosphere, or, taking the average melting point of rocky mixtures at 2,000° Fahr., to melt nearly three and a-half cubic miles of such rock, if of the same specific heat.

Of the heat annually lost by our globe and dissipated into space, represented by 777 cubic miles of ice melted, as before stated, the chief part is derived from the actual hypogeal source of a hotter though not necessarily fused nucleus, and nearly, if not wholly, is quite independent of the heat of Vulcanicity, which is developed as a consequence of its loss or dissipation. But were we to take the extreme case, and suppose it possible that all the heat the globe loses annually resulted from the transformation of the work of internal crushing of its shell, we shall find that the total volume of rock needed to be crushed in order to produce the required amount of lost heat is perfectly insignificant as compared with the volume of the globe itself, or that of its shell. For, as 1·270 cubic miles of crushed rock developes heat equivalent to that required to melt one cubic mile of ice to water at 32°, and if we assume the volume of our globe's *solid* crust to equal one-fourth of the total volume of the entire globe, 987 cubic miles of rock crushed annually would supply the whole of the heat dissipated in that time. But that is less than the *one sixty-five millionth* of the volume of the crust only.

But a very small portion of the total heat annually

lost by our globe is sufficient to account for the whole of the volcanic energy of every sort, including thermal waters, manifested annually upon our earth. In the absence of complete data, we can only approximately calculate what is the annual amount of present volcanic energy of our planet. This energy shows itself to us in three ways: 1. The heating or fusing of the ejected solid matters at volcanic vents. 2. The evolution of steam and other heated elastic fluids by which these are carried. 3. The work of raising through a certain height all the materials ejected. To which we must add a large allowance for waste, or thermal mechanical and chemical energy ineffectually dissipated in and above the vents. All these are measurable into units of heat.

I have applied this method of calculation to test the adequacy of the source I have assigned for volcanic heat, in two ways, viz.: 1. To the phenomena presented during the last two thousand years by Vesuvius, the best known Volcano in the world; and 2. To the whole of the four hundred and odd volcanic cones observed so far upon our globe, of which not more than one-half have ever been known in activity.

It is impossible here to refer to the details of the method or steps of these calculations. The result however is, that making large allowances for presumably defective data, *less than one-fourth* of the total telluric heat annually dissipated (as already stated in amount) is sufficient to account for the annual volcanic energy at present expended by our globe.

It is thus represented by the transformation into heat of the work of crushing about 247 cubic miles of (mean)

rock, a quantity so perfectly insignificant, as compared
with the volume of the globe itself, as to be absolutely
inappreciable in any way but by calculation; and as
its mechanical result is only the vertical transposition
transitorily of material within or upon our globe, the
proportion of the mass of which to the whole is equally
insignificant, so not likely in any way to produce changes
recognisable by the astronomer.

Space here forbids my entering at all upon that
branch of my investigation which is based upon the
experimental results, above mentioned, of the total contraction of fused rocks: for these, the original Paper can,
I hope, be hereafter referred to. I am enabled, however,
to prove thus how enormously more than needful has
been the store of energy dissipated since our globe was
wholly a melted mass, for the production, through the
contraction of its volume, of all the phenomena of
elevation and of Vulcanicity which its surface presents.
And how very small is the amount of that energy in a
unit of time as now operative, when compared with the
same at very remote epochs in our planet's history.

I have said that if we can find a true cause in Nature
for the origination of volcanic *heat*, all the other known
phenomena, at and about volcanic vents, become simple.
Lavas and all other solid ejecta of Volcanoes, from all
parts of the earth's surface, as well as basalts, present
in chemical and physical constitution close resemblance,
and may be all referred to the melting of more or less
fusible mixtures of siliceous crystalloid rocks with alu-
minous (slates, etc.) and calcareous rocks. Their general
chemical composition, and the higher or lower tempera-

tures of fusion resulting therefrom, together with the higher or lower temperatures to which they have been submitted at the different volcanic foci, determine their difference of flow (under like surface conditions) and of mineral character after ejection and cooling.

St. Clair de Ville and Fouqué have shown that the gaseous ejections, of which steam forms probably 99 per cent., are such as arise from water admitted to a *pre-existent focus of high temperature.*

Whether sea or fresh water is not material, when we bear in mind that the chemical constituents found in sea water and in natural fresh waters that have penetrated the soil are, on the whole, alike in kind and only differ in proportions. But I must pass almost without notice all the varied and instructive phenomena which are presented by volcanic vents, for to treat of these at all would be to more than double the size of this sketch.

In the source that has been pointed out as that from which volcanic heat itself is derived, viz., the secular cooling of our globe, and the effects of that upon its solid shell, we are enabled to point to that which is the surest test of the truth of any theory—that it not only enables us to account for all the phenomena, near or remote, but to predict them. We see here linked together as parts of one grand play of forces, those of contraction by cooling, producing by *direct* mechanical action the elevation of mountain chains, and by their *indirect* action, by transformation of mechanical work into heat, the production of Volcanoes ; and both by direct and by indirect action, of Earthquakes, never previously shown to have thus the physical connection of one common cause, but merely supposed, more or less, to be connected by their distribution upon our earth's surface.

We now discern thus the physical cause *why* Volcanoes are distributed, viewed largely, linearly, and follow the lines of elevation ; we see equally why their action is uncertain, non-periodic, fluctuating in intensity, with longer or shorter periods of repose, shifting in position, becoming extinct here, appearing in new activity or for the first time there. We have an adequate solution of the before inexplicable fact of their propinquity, and yet want of connection. We have an adequate cause for the fusion of rock at local points without resorting to the baseless hypothesis of perennial lakes of lava, etc.

For the first time, too, we discern a true physical cause for earthquake movement, where volcanic energy does not show itself. The crushing of the world's solid shell, whether thick or thin, goes on *per saltum* and at ever-shifting places, however steadily the tangential pressures producing it may act. Hence crushing *alone* may be shown to develope amply sufficient impulse to produce the most violent Earthquakes, whether they be or be not at a given place or time connected with volcanic outburst or possible injection, or with tangential pressures, enough still, in some cases, to produce partial permanent elevation.

When subterraneous crushing takes place, and the circumstances of the site do not permit the access of water, there may be Earthquake, but can be no Volcano ; where water is admitted, there may be both.

And thus we discern why there are comparatively few

submarine Volcanoes, the floor of the ocean being, on the whole, water-tight—"puddled," as an engineer would say, by the huge deposit of incoherent mud, etc., that covers most of it, and probably having a thicker crust beneath it than beneath the land.

We see, moreover, that the geological doctrine of absolute uniformity cannot be true as to Vulcanicity, any more than it can for any other energy in play in our world. Its development was greatest at its earliest stages, when the great masses of the mountain chains were elevated. It is even now—though as compared to men's experience, and even to all historic time, apparently uniform and always the same—a decaying energy.

The regimen of our planet as part of the Cosmos, which seems to some absolute (and presented to Playfair no trace of a beginning nor indication of an end), is not absolute, and only seems to us to be so because we see so little of it, and of its long perspective in time. This the now established doctrine of the conservation of energy renders certain.

With this source for volcanic heat, too, in our possession, we can look from our own world to others, and predict within certain limits, which must widen as our knowledge of the facts of their substance and surface becomes greater, what have been and what are the developments of Vulcanicity which have taken place or are occurring in or upon them. Looking to our own satellite, we see for the first time a sufficient physical cause for the enormous display of volcanic energy there which the telescope divulges to us; one which is not to be explained alone by the commonly made statement of the small density of the moon, but by the fact that as the rate of her cooling from a given temperature, as compared with that of our earth (apart from questions of the chemical nature of the two bodies, or of their specific heats, etc.), has been inversely as their respective masses, and directly as their surfaces, so has the rate of cooling of the moon been vastly greater than that of the earth, and the energy due to contraction by cooling more intense and rapidly developed in our satellite than upon our globe.

We have thus traced, in meagre and broken outline only—because space admitted no more—the progress of Science to its existing state as respects Vulcanicity, in its two branches of Vulcanology and of Seismology, and pointed out their more intimate relations and points of connection, and been at length able to refer them, on the sure basis of physical laws, to one common cause, and that one derived from no hypothesis, but simply from the postulate of our world as a terr-aqueous globe cooling in space.

What I have here advanced with reference to volcanic energy, which appertains to my own researches, I do not conceal from myself, nor from the reader, has yet to await the reception generally and the award of the true men of science of the world.

That, like every new line of thought which has attempted or succeeded in supplanting the old, it will meet with opposition, I make no doubt.

My belief, however, is that in the end it will be found to have added a fragment to the edifice of true knowledge.

The interpretation which I have given of the nature and origin of volcanic activity points at once to the function in the Cosmos which it is its destiny to fulfil. It is the instrument provided for the purpose of continually preserving the earth's solid shell in a state to follow down after the descending nucleus. It does this by an apparatus or play of mechanism whereby the material of the solid shell, locally or along certain lines, is not only crushed, but the crushed material is blown out as dust, or expelled as liquid rock from between the walls of the shell, which are thus enabled to approach each other; and thus, by relief of the tangential thrusts, to permit the shell to descend, which it is obvious that crushing alone, unless it extended to the whole mass of the shell, could not accomplish.

It is a wonderful example of Nature's mechanism thus to see how simple are the means by which this end is accomplished. The same inevitable crush that dislocates the solid shell along certain lines, produces the heat necessary to expel to the surface the material crushed.

When attempted to be made the basis for philosophic discovery, "final causes" are no doubt barren, as Bacon has said; but when we have independently and by strict methods arrived at a result, we may justly appeal, as a test of its truth, to its showing itself as plainly fulfilling a needful end, and, by a distinctly discernible mechanism, preserving that harmony and conservation which are the obvious law of the universe.

As has been said, if I mistake not by Daubeny, John Phillips, by Herschell, and by myself, the function of the

Earthquake and the Volcano is not destructive but, preservative. But we now see that: that the preservative scope of this function, as respects our earth, is far wider than what has been previously attributed to it. The Volcano does not merely throw up new fertile soil, and tend, in some small degree, to restore to the dry land the waste for ever going on by rain and sea; it fulfils a far weightier and more imperative task; it—by a mechanism the power of which is exactly balanced to the variable calls demanded of it, and which working almost imperceptibly, although in a manner however terrible its surface-action may at times appear to us little men*— prevents at longer intervals such sudden and unlooked-for paroxysms in the mass of our subsiding earth's shell as would be attended with wide-spread destruction to all that it inhabit.

To the popular mind, Volcanoes and Earthquakes are only isolated items of curiosity amongst " the wonders of the world:" few geologists even appear to realise how great and important are the relations of Vulcanicity to their science, viewed as a whole. Yet of Vulcanicity it is not too much to say, that in proportion as its nature and doctrines come to be known and understood as parts of the Cosmos, the nearer will it be seen to lie at the basis of all Physical Geology.

---

* " Magna ista quia parvi sumus"—SENECA, "Quæs. Nat."

END.

TRANSLATION

PROFESSOR PALMIERI'S

ACCOUNT OF

THE ERUPTION OF VESUVIUS

OF

1871—1872

I.

ACCOUNT OF THE ERUPTION.

THE great and disastrous conflagration of Vesuvius, which took place on the 26th of April, 1872, was, in my opinion, the last phase of an eruption which commenced at the end of January, 1871, an account of which I was unwilling to write, because I was convinced that it would not really terminate without a more or less violent explosion, such as I had often predicted. I shall now state the reasons upon which my prediction was founded.

When the central crater begins to heave, with slight eruptions, one may always predict a series of slight convulsions of greater or less duration, which are preparatory to the grand explosion, after which the Volcano remains for the most part in repose. Thus, when I observed the cone fissuring in November, 1868, and copious lava streams issuing from it, and flowing over the beautiful and fertile plains of the Novelle, through the Fossa della Vetrana, instead of announcing the beginning of an eruption, I announced the termination of one which had been manifest for upwards of a year by the constant flow of lava from the summit of the cone.

From the month of November, 1868, until the end of

December, 1870, the mountain remained quiet, except that the fumaroles at the head of the fissure showed a degree of activity by which chlorides and sulphides of copper, sulphide of potash and other products, were engendered.

But in the beginning of 1871 the seismograph was disturbed,(1) and the crater discharged, with a slight detonation, a few incandescent projectiles. Then I announced that *a new eruption had commenced, which might be of long duration, but with phases that could not possibly be foreseen ;* and on the 13th January, on the northern edge of the upper plain of the Vesuvian cone, an aperture appeared, from which at first a little lava issued, and then a small cone arose and threw out incandescent projectiles, with much smoke of a reddish colour, whilst the central crater continued to detonate more loudly and frequently. The lava-flow continued to increase until the beginning of March, without extending much beyond the base of the cone, although it had great mobility. In March, this little cone appeared not only to subside, but even partly to give way, as almost happens with eccentric cones when their activity is at an end. Upon visiting it, I observed that four prismatic or pillar-like masses remained standing, three of which were formed of scoriæ which had fallen back again in a pasty condition, and had become soldered together, the fourth consisting of a pyramidal block of compact and lithoidal lava, which appeared to have been forced up by impetus from the ground beneath. A little smoke issued from the small crater, and a loud hissing from the interior was audible.

By lying along the edge, I could see a cavity of cylindrical form about ten metres in depth, tapestried with stalactitic scoriæ covered with sublimations of various colours. The bottom of this crater was level, but in the centre a small cone of about two metres had formed, pointed in such a manner that it possessed but a very narrow opening at the apex, from which smoke issued with a hissing sound, and from which were spurted a few very small incandescent scoriæ. This little cone increased in size as well as activity until it filled the crater, and rose four or five metres above the brim.* New and more abundant lavas appeared near the base of this cone, and, pouring continually into the Atria del Cavallo, rushed into the Fossa della Vetrana in the direction of the Observatory and towards the Crocella, where they accumulated to such an extent as to cover the hill-side for a distance of about 300 metres; then turning below the Canteroni, they formed a hillock there without spreading much farther. These very leucitic lavas are capable of great extension, the pieces which are ejected forming for the most part very fine filiform masses, which may be collected on the mountain in great quantities, and specimens of which I presented to the Academy under the name of *filiform lapilli.* These threads were often of a clear yellowish colour, and, when observed under the microscope, were found to consist of very minute crystals of leucite embedded in a homogeneous paste.

---

* This small cone, as it appeared on the 1st April, is described and drawn in a Memoir of Professor von Rath, of the University of Bonn, on " Vesuvius on the 1st and 17th of April, 1871."

The crystals were still smaller as the diameter of the threads was less, and never formed knots or swellings even in the most hair-like threads. These observations led me to reject the opinion of those who hold that crystals of leucite are pre-existent in the lava. The viscous nature of these lavas prevented their being covered with fragmentary scoriæ, but caused the formation at first of a skin, which, thickening, became at last a more or less pliable shell, that, when more solidified, allowed the still fluid part to run as in a tube formed of this solid shell. For many months the lava descended thus from the cone and traversed the Atria del Cavallo, always covered, appearing below the Canteroni of a lively fluidity, until it could no longer be enveloped in its skin, which was stretched by the addition of new lava, and finally rent asunder to give room to the current until, owing to diminished liquidity, it was constrained to stop. When the lava, having traversed the covered channel it had made for itself from the top of the mountain to below the Canteroni, made its appearance still running, it frequently formed large bubbles on the surface, which mostly burst to give vent to smoke, and then disappeared.

In October, 1871, near the edge of the central crater, another small crater was formed by falling in, which, after a few days, gave vent to smoke and several jets of lava. The principal cone frequently opened in some point of the slope to give egress to small currents of lava, which quickly ceased. But towards the end of October the detonations increased, the smoke from the central crater issued more densely and mixed with

ashes, and the seismograph and accompanying apparatus were disturbed: for all these reasons, I said in
one of my bulletins, *we have either reached a new phase
or the end of the eruption*, not knowing whether the new
phase would be the last. On the 3rd and 4th November copious and splendid lava streams coursed down
the principal cone on its western side, but were soon
exhausted. The cone of 1871 appeared again at rest,
and partly even fell in, but did not cease to emit smoke
and to show fire in the interior.

In the beginning of January, 1872, the little cone
again became active, the crater of the preceding October
resumed strength, with frequent bellowings and projectiles, and soon after lavas of the same kind as before
reappeared. The cone of 1871, formed again by the
lava ejected, became so full that the lava poured from
its summit in the most singular and enchanting manner.
So far only an eccentric or ephemeral cone had risen
close to the central crater, which, after exhaustion,
regained vigour and discharged lava from the apex
instead of the base, as usually happens.

In the month of February matters were somewhat
moderated; but in March, with the full moon, the cone
opened on the north-west side—the cleavage being
manifest by a line of fumaroles—and a lava stream
issued from the lowest part without any noise and with
very little smoke, and poured down into the Atria del
Cavallo as far as the precipices of Monte di Somma.
This lava ceased flowing after a week, but the fumaroles
pointed out the cleft of the cone; and between the
small re-made cone, which had risen to the height of

35 metres, and the central crater, a new crater of small dimensions and interrupted activity opened.

On the 23rd April (another full moon) the Observatory instruments became agitated, the activity of the craters increased, and on the evening of the 24th splendid lavas descended the cone in various directions, attracting on the same night the visits of a great many strangers. All these lava streams were nearly exhausted on the morning of the 25th; only one remained, which issued from the base of the cone, not far from the spot whence that of the preceding month had issued. Numbers of visitors, attracted by the splendour of the lava streams of the preceding night, which they supposed still continued, soon arrived, but, finding them exhausted, were for the most part conducted by their guides to see the one still flowing. It was almost inaccessible, and to reach it one had to walk over the rough inequalities of the scoriæ. It took me two hours to get there from the Observatory, when I visited it that morning, and therefore I endeavoured to dissuade those who wished to visit it at night from the attempt, but set out myself from the Observatory at 7 p.m., leaving my only assistant there. The instruments were agitated. After midnight the Observatory was closed, and my assistant retired to rest. Late and unlucky visitors passed unobserved with an escort of inexperienced guides; at half-past 3 o'clock in the morning of the 26th they were in the Atria del Cavallo, when the Vesuvian cone became rent in a north-westerly direction, the fissure commencing at the little cone which disappeared, and extending to the Atria del Cavallo, whence a copious torrent of

lava issued. Two large craters formed at the summit
of the mountain, discharging numerous incandescent
projectiles with white ashes, and glittering with particles of mica, which frequently recurred.

A cloud of smoke enveloped these unfortunates, who
were under a hail of burning projectiles and close to
the lava torrent. Some were buried beneath it* and
disappeared for ever ; two dead bodies were picked up,
and eleven grievously injured, one of whom died close
to the Observatory. He alone revealed his name,
Antonio Giannone. I learned afterwards that he was
a fine young fellow, and Assistant-Professor in one of
the Universities.

Assistant-Professors Signor Franco, who is a priest,
and Signor Francesco Cozzolino, a priest also, entrusted
with the festive mass for the Observatory, hastened to
assist the dying. On my own return thither, the sad
spectacle of the dead and dying awaited me ; the former
were conveyed, through the assistance of the municipal
officer of Resina, to the Cemetery, and the latter to the
Hospital. But we must leave this scene of grief and
sorrow, and return to the eruption.

The fissure of the cone on the north-west side was
large and deep, and extended into the Atria del Cavallo,
about 300 metres. No mouth opened along the cleft of

---

* Eight young medical students perished beneath the lava, with
others unknown by name. They were all youths of good promise ;
their names will be recorded on the marble monument to be erected
near the Observatory. They are : Girolamo Pausini, Antonio and
Maurizio Fraggiacomo, Francesco Binetti da Molfettu, Giuseppe
Carbone da Bari, Francesco Spezzaferri da Trani, and Giovanni Busco
da Casamassima and Vitangelo Poli.

the cone itself; all the lava issued from that part which
extended into the Atria. From previous experience I
should have expected to have seen the formation of
adventitious cones along the widest part of the fissure,
which is never that most elevated, and these discharging
from their summits æriform matter frequently mixed
with projectiles, and from their base lava; but on this
occasion no cone appeared at the widest part of the
fissure, but a long hillock was formed like a little chain
of mountains, one point of which was elevated about
fifty metres above the plain beneath, and bearing no
resemblance to a cone.

Another fissure opened in the cone on the south side,
which did not extend to the base, and lava issued from
this and flowed in the direction of the Camaldoli.
Streams of less importance furrowed the cone in other
directions, but the largest quantity of lava proceeded
from the fissure in the Atria del Cavallo, below the
hillock or miniature chain of collines just described.
This lava stream was for some time restrained within
the Atria del Cavallo, among the holes and inequalities
of the lavas of 1871, but these being filled up and
overcome, it divided into two branches—the smaller one
flowing through a hollow which separated the lavas of
1867 from those of 1871, and made its way over the
lavas of 1858, threatening Resina, but stopped as soon as
it reached the first cultivated ground; the larger branch
precipitated itself into the Fossa della Vetrana, occupying the whole width, about 800 metres; and traversing
the entire length of 1,300 metres in three hours. It
dashed into the Fossa di Faraone; here it again divided

into two streams, one overlying the lava of 1868, on the Plain of the Novelle, partially covering the cultivated ground and country-houses; the other flowing on through the Fossa di Faraglione, over the lava of 1855, reached the villages of Massa and St. Sebastiano, covering a portion of the houses, and thence continued its course through the bed of a foss or trench which, contrary to my advice, had been excavated after the eruption of 1855, in the expectation of diverting the course of that lava. I did not fail to observe that the rains which previously descended through these steep channels, would in future be kept back to filtrate through the scoriæ, without ever reaching the new channel.

The lava of this eruption, meeting with this said excavation, flowed into it, instead of pursuing its road over the lava of 1855, and thus invaded highly cultivated ground and towns of considerable value, extending to the very walls of a country-house belonging to the celebrated painter, Luca Giordano. This lava stream, having surmounted the obstacles which the heaps of scoriæ in the Atria del Cavallo presented to it, ran with great velocity (notwithstanding its being greatly widened out in the Fossa del Vetrano), so that between 10 a.m. and 11 p.m. it traversed about five kilometres of road, occupying a surface of five to six square kilometres. If it had not greatly slackened after midnight, from the failure of supply at its source, in twenty-four hours more, by occupying Ponticelli, it would have reached Naples, and flowed into the sea.

Although I had often visited the two villages of

Massa and St. Sebastiano, previously greatly injured by the lava of 1855, yet I could not well estimate, upon now seeing them again, the number of houses which had disappeared. Massa seemed to me diminished by about one-third, and St. Sebastiano by somewhat less than a fourth. But the way of escape was open to the inhabitants of Massa; whilst a great river of lava occupying the road leading to St. Giorgio a Cremano would have hindered the flight of the inhabitants of St. Sebastiano, if they had been dilatory. The lava stream now separating the two villages is little less than a kilometre in width, and is about six metres in height.

On the night of the 26th April, the Observatory lay between two torrents of fire, which emitted an insufferable heat. The glass in the window-frames, especially on the Vetrana side, was hot and cracking, and a smell of scorching was perceptible in the rooms. The cone, besides being furrowed by the lava streams just described, was traversed by several others, which appeared and disappeared. It seemed completely perforated, and the lava oozed as it were through its whole surface. I cannot better express this phenomenon, than by saying that *Vesuvius sweated fire.* In the day-time, the cone appeared momentarily covered with white steam jets (fumaroles), which looked like flakes of cotton against the dark mountain-side, appearing and disappearing at brief intervals.

Simultaneously with the grand fissure of the cone, two large craters opened at the summit, discharging with a dreadful noise, audible at a great distance, an immense

cloud of smoke and ashes with bombs and flakes, rising
to the height of 1300 metres* above the brim of lava (*sull'
orlo de essi*). The white ashes, before described, although
they did not fall beyond the Crocella, were carried by
the wind as far as Cosenza, from whence they were sent
to me by Dr. Conti. These ejections were followed by
dark sand, with lapilli and small fragments of scoriæ of
the same colour. The smoke, driven up with violence,
assumed the usual aspect of a pine tree, of so sad a
colour that it reminded us of the shadowy elm of
Virgil's dreams (" *ulmus opaca ingens*"). From the trunk
and branches of the pine-tree cloud fell a rain of
incandescent material, which frequently covered all the
cone. The lapilli and the ashes were carried to greater
distances.

The victims of the morning of the 26th, the torrents
of fire which threatened Resina, Bosco and Torre
Annunziata, and which devastated the fertile country
of the Novelle, of Massa, St. Sebastiano and Cercola,
the two partially buried villages, the continual and
threatening growlings of the craters, caused such terror
that numbers fled from their dwellings near the mountain into Naples, and several in Naples went to Rome or
to other places. Very many delayed from the knowledge
that I was in the Observatory, and held themselves in
readiness for flight whenever I should abandon it.

---

* If this enormous height of projection really means, that above the
brim of the crater, it involves an initial velocity of projection of above
600 feet (British) per second.

Observations of the height of ascent of volcanic blocks are always
difficult and deceptive, and never free from error.—*Translator.*

The rapidity with which the vast torrent of fire
assailed the houses (*i.e.*, in these villages), and the great
heat which spread to a distance, scarcely allowed the
fugitives to carry away any of their belongings; many
were completely destitute. The authorities vied with
each other in zealous efforts to relieve the distress, and
the municipality of Naples sheltered and fed the wretched
beings for many days.

The igneous period of the eruption was short, for on
the morning of the 27th the lava stream, bearing down
upon Resina, having covered a few cultivated fields,
stopped; the lava descending from the summit of the
mountain towards the Camaldoli also stopped; and the
great lava torrent, which passed the shoulders of the
Observatory through the Fossa della Vetrana, lowered
the level of its surface below those of its two sides,
which appeared like two parallel ramparts above it.

If these streams had continued on the 27th, flowing
in the same manner as they did on the night of the
26th, they would have reached the sea, bringing
destruction to the very walls of Naples.

But before leaving the subject of these lavas I must
narrate an important fact to which I was witness, and
which was thrice repeated, near the banks of the great
river of fire that ran close to the Observatory. At three
several points, and at different times, I observed great
balls of black smoke issue from the lava, driven up with
continued violence, as if from a crater; through the
smoke I frequently observed numerous projectiles thrown
up into the air, but I could not say whether with noise
or in silence, for the noise of the central crater was

deafening. Each of these little eruptions, which I may call *external eruptions*, lasted from fifteen to twenty minutes. The first took place at the most elevated point of the Fossa della Vetrana, on the right bank of the torrent; the second, under the hill of Apicella, where the lava divided into the two branches, before described; and the third near to the Observatory on the left bank of the lava stream. These singular explosions terminated without leaving little cones or craters, the lava in its impetuosity carrying every trace away. These eruptions were seen from Naples, and the Observatory was justly believed to be in danger. One has been clearly photographed, the one which was the best seen from Naples, being the nearest and the least darkened by the smoke of the lava. (Plate 4.) Is this the first time that the phenomenon has been remarked? I believe that it is at least the first time it has been authenticated. The authority of Julius Schmidt, quoted by Scrope, has no weight with me, for I was also a witness of what happened at Vesuvius in 1855; and, although these cones were in the midst of the lava in the Atria del Cavallo, they originated, according to the opinion of everyone, from the fissure from which the other and much larger cones proceeded. The same phenomenon was observed in the Atria del Cavallo in 1858, when I caused two of the little cones to be brought to the Observatory; but these also might belong to the fissure along which the other cones were arranged. The same may be said of the little craters observed, after they had been exhausted, by Professor Scacchi in 1850. But the discharging mouths now observed in the Fossa della Vetrana, which existed for twenty minutes and then disappeared, and which were not at all in a continuous line, and could not be supposed to correspond with any fissure beneath, constitute a circumstance which, if not new, is evident for the first time, and cause the recognition of a power in the lava itself to form eruptive fumaroles.(2)

The igneous period of the eruption having terminated on the evening of the 27th, the ashes, lapilli, and projectiles became a little more abundant, whilst the roaring noises of the craters apparently became greater. The pine-tree cloud was of a darker colour, and was furrowed by continual lightning, visible by daylight from the Observatory. Many writers on the subject of Vesuvius affirm that the flashes which appear through the smoke cloud were lightning unaccompanied with thunder, but they studied the phenomena from Naples, or some place more or less distant from the crater, where the report of the thunder was inaudible, or could not be distinguished from the bellowing and detonation of the mountain. The fact is that these flashes were constantly followed by thunder, after an interval of about seven seconds.* When the flash was very short, a simple noise like the report of a gun was heard, but if it were long, a protracted sound like that from torn paper ensued.

---

* Assuming these flashes to have emanated from somewhere within the cloudy volume of steam and dust called "the head of the pine-tree," this interval would indicate that the mean height of this cloudy volume itself was not more than about four thousand feet above the top of the cone ; and, if so, that is not very far from the limit in height of projection of the dust and lapilli.—*Translator.*

On the 28th the ashes and lapilli, continuing to fall abundantly, darkened the air, yet without diminishing the terrible noise; at Resina, Portici, St. Giorgio a Cremano, Naples, etc., terror was universal.

On the 29th, with a strong wind blowing from the east, scoriæ of such a size fell at the Observatory, that the glass of the windows unprotected by external blinds was broken. The noise from the crater continued, but the projectiles rose to a less height, indicating a diminution in the dynamic power of the eruption. Towards midnight the noise of the craters was no longer continuous, and recurred with less force and for shorter intervals. Almost at the same hour a tempest burst over the Campania with loud thunder and a little rain. The grass, the seeds, the vine tendrils, the leaves and tops of the trees dried up immediately, and the country was changed from spring to winter. The storm, although repeated on the following days, passed away by degrees, and thus the floods, which I strongly feared, did not occur. Almost always after great eruptions of Vesuvius, storms of heavy rain have followed, and the ground being covered with ashes, the water could not filtrate through into the soil, but descended in muddy torrents over the adjacent country, occasioning as much damage as the fire itself.

On the 30th, the detonations were very few, and the smoke issued only at intervals, and by the 1st May the eruption was completely over.

When the smoke had cleared off the figure of the cone was seen to be changed. (*Vide* Plate 5A.)

The ground was perpetually disturbed whilst the

Volcano raged, so that the Observatory oscillated continually. Some shocks were felt not only in the adjacent territory, but at a greater distance, at Montovi and elsewhere. The oscillations at the Observatory were chiefly undulatory, from N.E. to S.W. They were observed for some days after the termination of the eruption, but not continuously, although they maintained some intensity.

If we refer to January, 1871, we shall find that that eruption was preceded by several earthquakes, among which were those of the months of October, November and December, in the previous year, that wrought such destruction in Calabria, and especially in the province of Cosenza; if we consider that as only the last phase, we shall find that it was preceded by great shocks of earthquake that devastated some regions of Greece.(3)

The great quantity of lapilli which fell buried the scoriæ with which the Vesuvius cone was covered, so that it became somewhat more difficult to ascend to the summit, and much less difficult to descend. Having reached the top of the mountain, I found a large crater divided into two parts by what seemed a cyclopean wall. The two abysses had vertical sides, and revealed the internal structure of the cone. Their vertical depth was 250 metres; and beyond that I observed a sort of tunnel perforated in the rock, with a covering arch raised above the bottom of the eastern abyss about 12 metres, judging by the eye. The interior walls of the crater showed neither the usual stalactitic scoriæ nor sublimations, nor fumaroles, but alternate beds of scoriæ and of compact lava. The fumaroles and sublimations abounded only about the brims of the craters. Hydrochloric and sulphuric acid and sometimes sulphuretted hydrogen affected respiration, and the temperature rose sometimes to 150 degrees. Various fissures about the brim of the double crater indicated prolongations downwards, which allowed me to descend with a rope, in order to examine the interior of the tunnel to which I have just alluded. The highest brim of the crater was fissured for a distance of 80 metres, and the greatest depth of fissure was at that place.

By measurement with the barometer, we ascertained approximately (for only one barometer was used) that the height of the Vesuvian cone was somewhat diminished.

Not only the Vesuvian cone, but the whole adjacent country appeared white for many days, as if covered with snow, when exposed to sunlight. This was due to the sea-salt contained in the ashes with which the surface was strewn.

A great quantity of coleoptera assembled on the flat roof of the Observatory, where the ashes and lapilli were heaped up two decimetres in height. I found the same species on the cone, where many insects were observed on other occasions, such as the *Cuccinella septempunctata;* the crysomela populi, etc., were wanting. This phenomenon of the extraordinary concourse of insects on the top of Vesuvius, in order to die in some of the fumaroles, especially noted previous to and after great eruptions, is a circumstance for which I cannot account. (4)   The whole of the lava emitted in this

eruption occupies a surface of about five square kilometres ; allowing an average thickness of four metres,
we obtain a mass of twenty millions of cubic metres.
About three-fifths of this lava did no injury, being deposited upon other pre-existing lava. However, the
lava in the Novelle, which was deposited upon the lava of
1858, covered quarries of the best stone which had been
worked at the time, covered many paths that had been
cleared, and buried the new Church of St. Michele,
with some houses that surrounded it, which had been
rebuilt on the site of the former church, which was
covered by the lava of 1868. The destruction of land
in occupation, of buildings and of crops, exceeded three
million francs in value. Many proposals for relieving
the sufferers have been received. Wishing to aid in
this benevolent work, I gave a public lecture, admission
for each person being one franc ; and this lecture, from
notes badly taken, was printed by private speculation,
and I was compelled to repudiate the report of it
through the public papers.

The evolutions of carbonic acid (*mofette*), which usually
appear at the end of great Vesuvian eruptions at low-
situated spots or hollows, with very rare exceptions,
were observed on this occasion a few days after the
eruption had completely ceased. They appeared in the
direction of Resina. I found the most elevated at Tironi,
and the most numerous between La Favorita and the
Bosco Reale di Portici.

The water in wells was on this occasion neither defi-
cient nor scarce previous to the eruption, but was very
acid after the appearance of the carbonic acid evolutions

in those neighbourhoods in which they abounded. Having stated that the disastrous conflagration of the 26th April ought, in my opinion, to be regarded as the last phase of a long period of eruption, which commenced at the beginning of 1871, I consider it right to discuss the question at somewhat greater length.

Not only from twenty years' personal observations, but from the attentive study of accounts of previous eruptions, I have found that when the central crater awakens with small eruptions after a certain time of previous repose, these almost always have a long duration, and, after various phases of increase and decrease, terminate in a great eccentric eruption, that is to say, with the production of an aperture from which a copious lava stream issues. The eruptions of 1858, 1861, 1868 and 1872, furnish the most recent examples of what I affirm. I might cite many others of earlier date, but I shall content myself with recording the greatest conflagration of this century, that of October, 1822.

Before the erection of the Vesuvian Observatory, it was impossible to obtain a consecutive account of all the phases which the Volcano presented ; but we generally obtained the description of the more splendid phases of the eruption which arrested the attention of everyone. Hence, notices of the small phenomena which preceded a great eruption are frequently wanting. We cannot always ascertain whether the fumaroles of the craters became active and at what periods, what was their temperature and what the diverse nature of their emanations, etc. : whether and when any change in the

H 2

crater with slight eruptive manifestations occurred; discharges which sometimes commenced in the bottom of a crater becoming active, and so are invisible at Naples.

But it may be asked whether the inverse proposition be equally true, that is, whether all the great eruptions of our Volcano were preceded by small fiery manifestations of long duration ?   There have undoubtedly been great eruptions not preceded by small central eruptions, but these also had their period of preparation or precursory signs.   After the great eruption of 1850, Vesuvius remained in apparent repose until the end of May, 1855, when there was an eccentric eruption and a great flow of lava lasting twenty-seven days.   But for a year before the fumaroles on the top of the mountain had acquired great activity, their temperature increased, and hydrochloric and sulphuric acid became more abundant, and generated the usual coloured products on the adjacent scoriæ.   Finally, in the month of January, a crater was formed by falling in of the ground, and although it did not discharge fire, yet it poured forth dense smoke. This was the beginning of the fissure manifested four months afterwards.

Ignazio Sorrentino, who spent a long life in the study of Vesuvius, and frequently ascended it, considered the increase of those yellow products—which are chiefly chlorides of iron, but were, at that time, mistaken for sulphur—as the sign of an approaching eruption.

The only grave objection that can be alleged is that of the memorable eruption of 1631, which surprised the neighbouring population so suddenly that many perished miserably, surrounded or covered with lava.   But that terrible conflagration occurred after centuries of repose, so that trees had grown in the interior of the crater. No one suspected the possibility of danger. It took place, too, at the end of autumn, when the cone is usually covered with clouds, and, therefore, no one had an opportunity of observing any precursory phenomena.

When the Observatory was established, I was able— in the first instance, at my own expense, and afterwards with some slight assistance from Government—to undertake studies more assiduous than any previously made. I had two instruments adjusted to indicate the internal efforts of the Volcano, viz., M. Lamond's apparatus of variations, which, by means of finely-balanced needles and methods of amplification proposed by Gauss, indicates the slightest trepidation of the ground, and my own electro-magnetic seismograph, a self-registering instrument of exquisite delicacy. These instruments, when attentively observed, give the most valuable information with respect to the activity of the adjacent Volcano.

If the very slightest eruption occurs, these instruments manifest slight perturbation, increasing with the activity of the mountain. When the Volcano attains a certain degree of activity, and the instruments are proportionately disturbed, it is impossible to foresee a new phase of increase without constantly watching the changes in the intensity of the perturbations; and to effect this it is requisite to have upon the spot a staff of assistants sufficiently numerous, scientific and intelligent. If, therefore, on the night preceding the 26th of April the instruments had been properly watched, they would have undoubtedly indicated the great increase in the activity of the Volcano. The perturbations on the 23rd were steadily increasing, and on the evening of the 25th they were much stronger than on the 24th, but on the morning of the 26th they had become extraordinarily strong; they must, therefore, have increased considerably during the night.

II

NATURE OF THE LAVAS.

WHEN the observer is near the source of the lava, he sees matter in a state of fusion, which, like a torrent of liquid fire, runs along, with more or less impetuosity, between two banks formed by itself. But as soon as the surface of the torrent cools to the point of congelation, it loses the splendour of its first incandescence. The part which begins to harden breaks readily in some lavas into fragments which float on the viscous fluid beneath; these, increasing in number with distance from the source, conceal the molten matter beneath and retard its progress, and at last nothing is seen but the more or less red-hot scoriæ moving along. These lavas I shall call "*Lavas with fragmentary scoriæ.*"

On other occasions, a skin forms on the surface of the lava, which, gradually thickening, keeps flexible for some time, and then wrinkles or swells or extends and breaks to give egress to the hot fluid within, which, in its turn, skins over and repeats the same phenomena. This I shall call "*Lavas with a united surface.*"

These, in their course, discharge less smoke than the first, draw out more easily into threads, and, when cold, have a dark colour, something like bitumen or pitch.

The lava with fragmentary scoriæ, when stretched, breaks easily, discharges smoke copiously, and, when hardened, has a more bluish tint, like clods of upturned earth (*formato di zolle*). It is noisy in its course, because the incoherent scoriæ that it carries along strike and crunch against each other; the other lava flows silently, except for a sort of crackling arising from the actual fracturing up of the solid skin by distension from the liquid matter within. If required to give the mineralogical characteristics of this lava, I would say that it was rich in leucite and contained little or no pyroxene; the fragmentary lava, on the contrary, is poor in leucite and rich in pyroxene. The lavas of 1871 were of the " united surface " character; those of 1872 were " fragmentary," with some characteristics which I shall describe:

1. They were of the clearest tint I have ever seen, when regarded superficially, but, when broken, the fracture was darker than any other lava.

2. They had very little leucite and abounded in pyroxene and olivine, and sometimes contained a few crystals of amphibole.

3. Their specific gravity varied with their porosity; the most compact attained 2·75.

4. These lavas carried along in their course a quantity of scoriæ which had long been subjected to the action of the acids of the fumaroles close to the craters, and also a great many bombs (*bombe*)—that is, round masses similar to those ejected from craters. These varied in

size, some having a diameter of four to five meters. They frequently contained a large nucleus of very leucitic lava, like that of 1871, with a larger or smaller quantity of feroligiste (peroxide of iron). Others contained lavas changed by the action of the acid vapours near the craters. These bombs must have flowed out with the lava, for they are found through its whole course, and they were certainly not ejected from the crater ; for not only are they found on the lava exclusively, but masses so enormous were not thrown up from the craters during the eruption ; those lying on the cone near the craters seldom exceed a decimetre in diameter.

As to the qualitative chemical analysis of the lavas, it always presents the same elements, with the exception of small quantities of some metals, lead for example, which have escaped the researches of good chemists, but which I have constantly found in the sublimations of the fumaroles of the lava. With respect to the quantitative analysis, two specimens of the same lava appear indeed to have their constituents in different proportions. To arrive at any conclusion a long and patient investigation, requiring means and assistance which the Observatory does not possess, would be necessary.

Professor Fuchs, of Heidelberg, has devoted himself to this work for years past, and if he continue it with well-selected and sufficiently large specimens we may hope some day to obtain satisfactory results.

5. Every specimen of lava which I examined with
a very sensitive magnetoscope improved by
myself, was invariably magneto-polar, not ex-
cepting the pieces of the bombs, whether rejected from the crater or carried along with the
lava.

III

FUMAROLES OF THE LAVAS.

SMOKE generally issues from all lava when it cools down to a certain degree, hence it is more abundant at the edges of the fiery torrent, or is liberated from the scoriæ that form on its surface. But when the lava stops, the smoke issues only from certain vent-holes, through which we can still see the fire, and at the edge of which different amorphous or crystallized matters collect by sublimation. These centres of heat, of more or less duration, are the fumaroles of the lavas. I believe I have on other occasions shown that a fumarole is nothing but a communication between the more or less cooled and hardened surface of the lava and the interior, which is still incandescent. Some fumaroles last but a day, others preserve their activity for weeks, months or years, according to the depth of lava through which they penetrate; and when they cease to be active, that is, when the sublimations are formed, or smoke or other æriform matters issue from them, they still retain a rather elevated temperature. In the lavas of 1858, in a place where they had a transverse width of 150 metres, a vent-hole may still be found where the thermometer registers $60°$ and the scoriæ are warm. Some-

times, while the lava is in process of cooling, new
fumaroles appear, in which the fire is visible. This
phenomenon, which appeared marvellous and inexplicable when I first observed it in 1855, is now very
easily understood ; the cooled and hardened crust of the
lava fractures with noise and suddenly, and so a new
communication is opened with the incandescent lava
below, thus creating a new fumarole.

As the smoke of the fluid lava is perfectly neutral,
that is, neither acid nor alkaline, so the fumaroles at the
first period of their existence with sublimations of sea-
salt, mixed frequently with oxide of copper either in
black powder or in shining laminæ, ought also to be
neutral. But if the fumarole continues active, hydrochloric acid issues with the smoke, and often some time
after sulphuric acid. Then the sublimations turn first
yellow, then green, and more rarely azure. The chemical reactions show that these sublimations are chlorides or sulpho-chlorides, and sometimes sulphides, and
they afford reactions, indicative of soda, magnesia,
copper, lead, and traces of other substances, not excluding ammonia, which I must speak of separately. This,
I have observed, is the general law with the fumaroles
of the tranquil lavas, which occur with long and moderate eruptions—for instance, the lavas of 1871, and even
those of 1872, preceding the 26th April.

But in the great lavas of the great conflagrations of
Vesuvius, chloride of iron more or less in combination
with all the other substances above mentioned changes
the appearance of the sublimations. The fumaroles in
the lava of the 26th April frequently indicated chloride

of iron. Sulphuretted hydrogen, by reaction of sulphurous acid, is decomposed, and sulphur sublimed, having a particular aspect, collects on the scoriæ. This is never found but in fumaroles of the smaller lavas ; it was therefore absent in those of 1871, but frequently occurred in those of 1872.

Although the sublimations are generally mixtures, yet sometimes distinct and crystallized chemical or mineral species are found, such as sulphur, sal ammoniac, *tenorite*, *cotunuite*, etc. Micaceous peroxide of iron (feroligiste), so common near eruptive cones, is very scarce on lava ; any found in it has been carried down from the craters, and proofs of this transport are very abundant and striking in the lavas of this last eruption. Even the iron found in the bombs is evidently transported ; there is a fumarole on the ridge of the lava in the Fossa di Faraone which contains micaceous peroxide of iron, and this, at first sight, appears to oppose what I have affirmed ; nevertheless, it gives additional force to my statement. This fumarole is only a bomb or rounded mass of enormous size, four or five metres in diameter. Smoke and hydrochloric acid issued from the aperture in its envelope, and being partly broken it was seen to contain lapilli and pieces of antecedent lava, covered with micaceous peroxide of iron. The internal temperature of this mass was very high ; the hydrochloride acid which it discharged had, in some places, covered the micaceous iron with a yellow coating of chloride of iron. From small apertures, on the lower side of the mass, white and green stalactites of chloride of calcium were visible. In one spot only of lava I found a fumarole, with a small quantity of micaceous peroxide of iron, evidently in a state of formation; but this was the very spot where the lava became eruptive, and whence issued the column of smoke which was so well photographed—the place under the hill of Apicella. (See Plate 4A.)

I have enumerated the products which are constantly collected in fumaroles, although they are not all found at the same time or place, in order to show that the sublimations follow a certain law in their appearance. *Tenorite*, for instance, was formerly considered an accidental product of certain eruptions, and I have always found it; but if you visit the fumarole when the acids have had time to transform it, you will no longer see it. I found the crystallized chloride of lead, or " cotunuite," as it is called, for the first time in the lavas of 1855, and thought it a singular circumstance; but from that time I recognised it in all the lavas, though not always so beautiful and abundant; and even when not found as a distinct substance, I observed it in combination with chloride of copper. In the lavas of the 26th April *cotunuite* and *tenorite* * were not very abundant, because the chloride of iron disturbed the greater number of the sublimations. I found sal ammoniac very abundantly on the fumaroles of the lavas that invaded the cultivated ground. Although chloride of ammonia, contrary to

---

* COTUNUITE, chloride of lead, in white, lustrous, acicular crystals, of the trimetric system, easily scratched, Sp. gr., 5·238.

TENORITE, peroxide of copper, in thin, hexagonal plates or scales, translucent when very thin, dark steel gray, of the cubic system; hard and lustrous. Sp. gr. about 5·950.—*Translator.*

opinion, was not wanting in the sublimations of the fumaroles of the lavas deposited on other lavas, yet it was neither abundant nor crystallized, but combined in small quantities with other substances. It appeared in great abundance in all the fumaroles of lavas which covered cultivated or woody ground. At first it was scarce enough, and mixed with chloride of sodium ; but when the rains came the sea-salt was washed away, and sal ammoniac formed beautiful crystals, nearly free from adventitious matters, as was the case with the fumaroles of the last lava. Afterwards, when chloride of iron was produced, ferro-chloride of ammonia was found. Crystals of sal ammoniac were sometimes found of a beautiful amber yellow. This colour was, in the opinion of my colleague, Professor Scacchi, produced by such small traces of chloride of iron that neither Professor Guiscardi nor I, nor indeed any other chemists to whom I submitted specimens for examination, could detect any. What I can affirm with certainty is, that these limpid crystals of a yellow colour were almost always attached to an amorphous substance, soluble in water, composed of various chlorides, in which iron was often detected.

From these remarks, it is evident that in the tranquil lavas the sublimations appear with a certain order of succession, and in the violent lavas, and those which flow most copiously, they are more complicated, and render both chemical analysis and spectroscopic researches more difficult. Notwithstanding, I observed traces of lithium and thallium, which I had previously perceived in some sublimations of 1871. I purpose

submitting many sublimations which I have collected
to more complete spectroscopic investigation, although I
am persuaded that the discovery of traces of certain
bodies in the sublimations or in the lavas is a matter of
small importance to the science of volcanoes. I must
say, however, that calcium was discovered on this occasion in great abundance, not only by the spectroscope,
but also by chemical analysis. Sulphate of lime has
often been found in larger or smaller proportions, but
this was the first time I had observed chloride of calcium
both close to the craters, and also in the sublimations of
the fumaroles upon the lavas. The white stalactites
which I collected beneath the great mass or bomb
above described were almost exclusively composed of
chloride of calcium, and only a few green drops
manifested, with the usual re-agents, the presence of
iron.

I did not fail to look often at the spectrum of the
flowing lavas covered with the smoke which issued from
them, but I always had a continuous spectrum. The
spectroscope employed was Hoffmann's construction,
with direct vision; but I think it would be better on
other occasions to use a spectroscope combined with a
telescope, like those used by astronomers.

But avoiding minute particulars of these sublimates,
let us see what is the general direction and the order of
their appearance. Sublimations are generally oxides,
chlorides and sulphates, sometimes sulphides. Among
the oxides, we must enumerate in the first place "tenorite" and *feroligiste* or micaceous peroxide of iron. The
first is almost always found at the commencement of

activity in the fumaroles, simultaneously with the sublimation of chloride of sodium; the second—which is, perhaps, never wanting in eruptive cones that are often found lined with it inside—is seldom generated in the fumaroles of the lava, and therefore it is not easy to define the moment of its appearance. Sometimes one collects micaceous peroxide of iron on the lava, but it is often transported there from the mouths of eruption, as happened on this occasion.

Trustworthy writers are of opinion that all the oxides are derived from the decomposition of the chlorides, but I think I have clearly demonstrated that, with regard to copper and lead, the opposite statement may be affirmed; for the oxides are changed into chlorides, and hydrochloric acid liberated. Oxide of copper forms sublimates at the beginning, at the same time as the sea-salt; and if the fumarole be anhydrous or, as Deville would say, *dry*, this oxide does not change into either a chloride or a sulphate; but if the fumarole gives watery vapour, after a little hydrochloric acid is formed, which changes the oxide into a chloride, and if whilst this is going on oxide of lead be developed, it is changed into the chloride of lead, so frequently found in combination with chloride of copper. Then the sublimations change from white to red or yellow, and specimens when carried away gradually turn light blue, but when heated on platinum over a spirit lamp they resume their yellow tint. Sometimes the yellow colour remains longer, and in time changes to green; this also happens on the fumarole itself, the green commencing at the zones furthest removed from the centre, where the temperature is

I

highest. When these sublimations are greenish, they become far less soluble than at first. The yellow, so common at a certain period on the fumaroles of the tranquil lavas, never attracted attention before I first examined it, doubtless, because it was considered chloride of iron, and yet in small eruptions this is only found close to the discharging mouths, and never in the sublimations of the fumaroles of the lava; but, on the other hand, it is the most copious and common product on the lavas of the great eruptions. This probably also accounts for the fact that lead, which is so obvious in the fumaroles of the lavas, had never previously been observed. In 1855, I noticed the crystallized chloride of lead in a fumarole in the Fossa della Vetrana, and this induced me always to look for it on the fumaroles of the later lavas; and I ascertained that, if it did not always appear as a distinct mineral, it was easily discovered in combination with other chlorides. The specimens which I have collected are not the most beautiful, but the presence of lead in the sublimations is not less common.

Micaceous peroxide of iron, when found on the lava, has been mostly conveyed from the eruptive mouths, as I have already stated, and perhaps never so abundantly and evidently as on this occasion. The lava of the 26th of April carried along a large quantity of round masses or bombs, varying in size, among which were found antecedent lava more or less covered with micaceous iron, either collected in the cavities of the lava, or incorporated with its mass. Sometimes the micaceous iron appears like little veins in the paste of new lava enveloping the exterior of these rounded masses, an exterior compact and lithoidal, and not resembling scoriæ. Among these spherical masses I found one of enormous size, four to five metres in diameter, which, having broken up where the exterior envelope was thinnest, I found filled with a great mass of lapilli and fragments of other lavas covered with micaceous iron. This bomb still preserves (June 5th) an elevated temperature within, and emits smoke and hydrochloric acid, which, meeting the micaceous iron discovered by breaking the envelope with blows of a hammer, transforms it superficially into chloride of iron, showing most clearly how, on some occasions at least, chloride of iron is formed from the oxide which precedes it. That those lapilli and the pieces of lava were solid when enveloped in the paste of the new lava, we infer from seeing the impressions on the inside of the said envelope. The chloride of calcium, which I found in this spherical mass almost pure, caused me to suspect that the sulphate of lime which is so often found on Vesuvius is a transformation of the chloride produced by the contact of sulphurous acid, which easily becomes transformed into sulphuric acid. The hydrochloric acid which escapes from a fumarole coming into contact with the scoriæ near its mouth, produces chloride of iron, which is, therefore, not always obtained by sublimation, although, when the temperature is very high, chloride of iron is conveyed from the interior of the lava, and sublimes on the exterior and colder parts; for instance, the chloride of iron which issues from the eruptive cones is sometimes found sublimed on the rocks of Monte di

Somma. When chloride of iron has been produced by sublimation, we may collect it inside a glass bell placed over the fumarole, or upon a piece of brick; but when it is produced by the action of hydrochloric acid on the scoriæ, it will only be found on the scoriæ themselves.

If, therefore, the origin of micaceous peroxide of iron were due to the decomposition of the sesqui-chloride of iron requiring a more elevated temperature for its decomposition, it would follow that its genesis would be easier near the discharging mouths, and more difficult on the lavas, but there the fact was verified: for example, in the great bomb on the fumarole, where we observed micaceous iron transformed into chloride of iron. We may therefore consider it *proved* that some chlorides—for instance, chloride of sodium—issue from the lava itself, either being there pre-existent, or being formed there; and that others are derived from the oxides which precede them, as undoubtedly is the case with chloride of copper; hence, the theory that derives the oxides always from the chlorides cannot be considered true. Granting that this theory might be applicable to the origin of micaceous iron, we should still want to know how it is found with the paste of the new lava itself, which forms the exterior coating of the bombs above described.

Many of these rounded masses, which have been rolled along by the lava, contain scoriæ partly decomposed by the long action of the acids found on the fumaroles of the craters. They disintegrate easily, and have a more or less yellowish tint. In the greater number of cases the interior of these masses is formed of leucitic lava, with cavities lined with micaceous iron. In short, their contents appeared to me quite similar to the material of the cone of 1871 and 1872, which in all probability was engulfed in the large crevasse or fissure that opened below it; and the fragments having thus fallen down into the lava, were enveloped by it and carried out by it after having been more or less rounded. The external envelope of these spheres is not at all scoriaceous, but compact and lithoidal, and sometimes composed of concentric folds or plaits.

As to the gaseous emanations of fumaroles, watery vapour with few exceptions comes first; this conveys the material which first appears in the sublimations, viz., sea-salt, and for the most part oxide of copper. If the fumarole continue active, it passes from the neutral period to the acid period, and first hydrochloric acid is produced, which, in small lava streams, never conveys chloride of iron, and rarely attacks the scoriæ to form that salt, but expends its force in changing the sublimations already there. For this reason chloride of iron, though completely absent in the lavas of 1871, was abundantly found in those of the 26th April, 1872. Sulphurous acid follows hydrochloric at a later period, and sulphuretted hydrogen occasionally succeeds.

Having examined the gases of fumaroles by means of a graduated tube, and the pyrogallate of potash, I always found that it contained less oxygen than the surrounding atmosphere.

For several years I wished to see whether the fumaroles of the lavas had a period of evolution of carbonic acid, as sometimes happens with fumaroles near the craters, but

I have always obtained negative results.   I often found
that the atmosphere on the lavas contained an excess
of carbonic acid, but as these lavas had burnt many
trees, and it was probable that carbonic acid springs
had formed under the lava, I never considered it safe to
form any conclusion on the subject.

IV

BOMBS, LAPILLI AND ASHES.

THE bombs ejected from the craters are like those carried down by the lavas, but of smaller size, and they seldomer contain a nucleus similar to those found in the latter. With the bombs properly so called, many pieces of incandescent lava were thrown up, and in their fall went beyond the base of the cone. A quantity of small scoriæ varying in size accompanied these projectiles, and those fragments, which we call *lapilli*, fell at a greater distance. With the lapilli, and sometimes without them, the smoke carried a very minute dust or sand, which is generally called ashes. These ashes, when washed with water, lose soluble constituents which they have collected in the smoke—such as chloride of sodium and other chlorides and often free acids. The insoluble part originates in the detritus of lava, and with the microscope we can detect abundant fragments of those crystals which most frequently occur in the lava of the same eruption. ·

The lavas of 1871, which were eminently leucitic, and almost entirely deprived of pyroxene, resembled the ashes, which appeared to be fragments of crystals of leucite, more or less enveloped in the paste of the lava, so that having triturated the scoriæ of the lava, and looked at the powder through the microscope, it was apparently quite the same as the ashes.

But at the beginning of the eruption of the 26th
April, a white sand fell in the Atria del Cavallo, close
to the Crocella (5), which on the dark scoriæ of 1871
looked like snow.   Its fall had a limit so well defined
that one passed without any gradation from white to
black.   Having collected some of this sand that very
morning, I put it up in white paper, for at that moment
it was impossible for me to examine it.   Taking it out
some days after, I found it had become reddish, and
having put it under the microscope, I observed that it was
exclusively formed of little pebbles more or less round,
of a transparent vitreous matter, partly covered with a
red substance.   Fragments of green crystals occurred
in this sand, upon which no red was perceptible.   I
consulted our eminent crystallographer, Arcangelo
Scacchi, whether these little pebbles were leucite, as I
suspected, and whether the green particles were pyroxene: he confirmed my suspicion, and remarked that
the red colour was superficial only.   We then washed a
little of the sand in hot water, and saw the pebbles
become whitish; but having heated some on platinum,
we observed that they first turned black and then became perfectly white, proving that the red was a deposit
of organic matter.   To see these leucites, rounded like
small pebbles transported by a torrent, deprived of the
soluble chlorides which generally accompany Vesuvian
ashes, is a matter worthy of attention.   Whilst heating
this sand upon platinum, decrepitation was audible,
which indicated the cracking of some of the little
pebbles.   It is evident, therefore, that crystals of leucite
raised to a certain temperature may break, and thus
we can understand how almost all Vesuvian ashes con-

tain fragments of the said crystals enveloped in the paste of the lava. It is evident that the soluble part of the ashes is obtained from the smoke through which it passes. On this occasion the smoke from the craters did not apparently contain much acids, for no bad smell was perceptible, and the water in which I washed the ashes scarcely reddened litmus paper. Even chloride of iron, which was so abundant in the lavas, was scarcely perceptible in the smoke, which almost exclusively deposited sea-salt on the surrounding rocks ; I say sea-salt advisedly, and not chloride of sodium, to show that I include all that sea-salt contains. The slight disturbance it manifested with chloride of barium, and the small precipitate with oxalate of ammonia, reveal sulphate of lime, without excluding the possibility of the chloride.

But how can these ashes do so much injury to the vegetation of the ground they cover, especially at the first fall of rain? I think that the damage is due partly to the sea-salt, and partly to the acids contained either in the ashes or in the rain-water itself. Upon watering the tender tops of some plants with a saturated solution of the salt from Vesuvius itself, I noticed that they withered away after a few hours. But very often the rain alone which traverses the smoke of Vesuvius, or is produced by condensation from it, gives manifest acid reactions, and destroys the grass and the tops of the trees. The peasants believe that the rain is warm or of boiling water, from observing that the tender parts of the plants are, by its deposit, all burnt up. Vegetation is now recovering, but without flowers, and consequently without fruit.

V

THE CRATERS AND THEIR FUMAROLES.

THE greater part of the lava issued from the base of the
great fissure in the cone which I have described; and
although two other lava streams descended from the top
of the mountain, neither proceeded from the crater, but
from apertures near it. The great crater, divided in two
as already described, opened wide on the morning of the
26th April, destroying the brim of the antecedent crater,
and remaking it in another shape with ejected matter,
except on the south-west side, where the brim was split.
(See Plate 5.)

From this double crater, copious smoke, bombs and
incandescent scoriæ, with ashes and lapilli, issued with
violence, and from the depths below came dreadful
detonations and bellowings, producing great terror.
And yet the lava poured out into the Atria del Cavallo
without any noise, and not even a column of smoke
marked its origin of issue—namely, from the fissure.

When the eruption was over, the sight of the vertical
walls of these deep craters, of almost horizontal
strata of scoriæ and lithoidal masses, with a fracture
fresh, and as if they had never undergone the action of

fire or of acid vapours, without recent scoriæ and without
fumaroles, was to me a marvellous spectacle. The
fumaroles were almost all on the brims of the craters,
with emanations of hydrochloric and sulphurous acid.
In a few that were more removed from the brim, sulphuretted hydrogen was perceptible. In the sublimations, chloride of iron was most abundant, in combination with other chlorides, for example, of sodium,
magnesium and calcium. This last chloride was frequent even among the sublimations of the fumaroles of
the lavas, and it was the first time it was ever remarked,
but I do not think it was the first time that it was
ever produced: being in combination with chloride of
iron, and very deliquescent, it did not attract attention
from anyone. In a hollow fragment of scoriæ I observed a yellowish substance, which looked like sulphur
in a viscid state, and which boiled at a temperature of
120°, and evolved hydrochloric acid. Having collected
this substance and poured it into a glass phial, it quickly
coagulated into an amorphous mass of the same colour;
but before I reached the Observatory, I found that it
had become liquid by deliquescence. It consisted of a
mixture of the aforesaid chlorides, according to an analysis made by Professor Silvestro Zinno and myself. In
some fumaroles, where I perceived the smell of sulphuretted hydrogen, I found sublimed sulphur under
the scoriæ.

At the source of the lava stream that flowed towards
the Camaldoli, on the seaward flank of Vesuvius, I
observed large fumaroles of steam only, pure aqueous
vapour.

There was no trace of carbonic acid in these fumaroles, but that fact does not imply that there was none
at a later period, for, since the first investigations of
Deville, it is known that carbonic acid is found under
certain conditions on the very summit of Vesuvius.

VI

THE ELECTRICITY OF THE SMOKE AND ASHES.

OUR ancestors could judge that a great amount of electricity was occasionally evolved in the smoke, from their observation of the lightning flashes that darted through the Vesuvian pine tree; but they had no proper instruments for ascertaining whether this evolution of electricity was constant or accidental, or what laws regulated its manifestations. My *apparatus, with movable conductor*, by which comparative observations of electric meteorology can be made, and the errors arising from dispersion corrected, supplied me with an easy method of studying the electricity evolved during eruptions.

I must begin by describing the bifilar electrometer, in order to explain the apparatus which I have named as above, "*Apparechio a conduttore mobile.*"

A A (Plate VIA, Fig. 1) is a glass cylinder, the lower edge of which is ground, well varnished with gum lac, and let into a wooden base, *B*, furnished with three levelling screws. Through a sufficiently wide glass tube, *a a*, runs a copper rod covered with insulating mastic, having a little plate or cylindrical cavity of gilded brass at the top (Figs. 2 and 3), with two arms dd, $d'd$. In the plate a disc of aluminium, m, is suspended by means of two silk fibres, and to the disc a very fine aluminium wire is attached, ff', bent a little at the ends, as are the arms, dd, $d'd$. The disc has about three millimetres less diameter than the plate. The diameter of the plate may vary within certain limits, but I have found it convenient to make it eighteen millimetres. The glass tube, aa (Fig 1), should descend below the base as much as it rises above it, that is three to four centimetres. The length of the index is about one decimetre.

The upper ends of the two silk fibres, by which the disk and index are suspended, are attached to the top of the glass tube, C, by a contrivance which permits a change in the distance between the two points of suspension, and a screw, p, is provided to raise and lower the disc with the index. At n, at the lower part of the tube, C, there is a kind of torsion micrometer, arranged so as to bring the index to the zero of the scale engraved on the graduated ring, B, which is formed of a strip of good paper pasted on the rim of a glass disc. The index must be placed at the zero of the scale, and must be some distance from the ends of the arms of the plate with which it is parallel. The plate is about three millimetres deep.

Having levelled the instrument, so as to render the disc concentric with the plate, and placed the index at zero, it is obvious that if an electric charge through the wire, h, reach the plate with the arms, it will electrify

the disc and index : the disc will have the opposite
electricity, and the extremities of the index will take
the same electricity as the arms, and consequently the
index will describe an arc more or less great.   The
motion of the index is sufficiently slow to allow the eye
conveniently to follow it.   Having traversed the first
arc, which I call the *impulsive* one, the index returns,
and, after only two oscillations, comes to rest at what I
shall call the *definite* arc.

When the electric charges are of very brief duration,
the impulsive arcs are within certain limits proportional
to the tensions, and the ratio between the impulsive and
definite arcs is expressed by the following equation :

$$\frac{\alpha\,(\beta - \alpha)}{\beta} = \text{tang. } \tfrac{1}{2}\,\alpha$$

In which β is the impulsive arc and α the definite arc,
showing that α comes out nearly equal to $\tfrac{1}{2}\,\beta$.   In dry
weather all goes perfectly within the limits of propor-
tion, and I can tell whether, during the time in which
the index traversed the impulsive arc, there were any
dispersions and of what nature ; for if the definite arc is
not close to the limit of the impulsive arc, it is a sign of
dispersions having taken place during the motions of the
index.   Every degree less in the definite arc denotes two
degrees of loss for the impulsive arc ; but as the index
employs double the time traversing the definite as it
does the impulsive arc, we may consider the loss of one
equal to the loss of the other.

In excessively damp weather the index gives no definite arc, and it is necessary to resort to artificial heat in
order to dry the insulators.   The most simple means I

know of is to hold the instrument over some hollow vessel, which, for the time, is converted into a stove by the introduction of a spirit lamp.

From Gauss's formula for the bifilar system of instruments of this class, we learn that the maximum sensitiveness of such instruments is given when the length of the suspending fibres is greatest, and the distance between them is smallest, with the weight of the movable or rotating member a minimum ; and these elements being the same, the sensitiveness of the instruments is invariable.

To some electrometers, in order to avoid errors of parallax, a small telescope, with a micrometer wire, has been added ; but, with a little practice, we can read accurately without this refinement. In order to obtain comparative measurements, it is necessary to select some given unit of tension. I have observed that by making a galvanic pile of copper, zinc and distilled water, and insulating it well, each pole has a tension which remains the same for many days, if the conditions of temperature and the moisture of the surrounding atmosphere are not very different. With thirty pairs of this pile, each element having twenty-five square centimetres of surface, I have on the electrometer a definite arc of 15°, with the temperature of the atmosphere at 20° C., and with the difference of 4° to 5° C. between the thermometers of the psychrometer of August's construction. The first observation was made twenty-four hours after mounting the pile. For unit of tension I took that which corresponded to a single pair, that is, the thirtieth part of the total tension. Other electrometers may be compared with one already properly adjusted, without always having recourse to the pile.

This done, let us see the arrangement of all the apparatus:

H H (Plate VIIᴀ, Fig. 1) is the ceiling of a well-situated lofty room, with an opening, *o o*, at the upper part.

M M, a bracket or table fastened against the wall, about a metre distant from the ceiling, *H H.*

N N, a wooden platform for the observer.

A, the bifilar electrometer.

B, Bohnenberger's electroscope.

a a, a movable conductor formed of a brass rod 15 to 18 millimetres in diameter, insulated below by means of a glass rod, well varnished with gum lac, having a suspending pulley, *c*, and a wooden guide-rod underneath it, *l*, within the guiding tube, *k*. At the upper part of this conductor, *a a*, there is a sliding roof, *b*, which can be adjusted so as to prevent rain entering at the opening, *o o*. The conductor terminates in a disc made of a sheet of thin brass, *d*, 24 centimetres in diameter. Upon this disc, or even in place of it, we may use metallic points.

As a support to the conductor at the upper part, I have made use of a triangular ring, *x*, drawn at its full size in Fig. 2. The conductor passes between three springs, and the triangular ring is held in place by three silk cords, *m m m*. Their material should not be mixed with any cotton, and it may be advisable to saturate them with an alcoholic solution of gum lac.

K

fff is a hempen cord, which is used to raise and lower the conductor.

i is a copper wire covered with silk, by means of which the triangular ring, x, and through that and its springs the conductor communicates with either the electrometer or the electroscope.

Quickly raising the conductor by pulling the cord, f, the index of the electrometer will describe a more or less large impulsive arc, and, after two oscillations, will stop at the definite arc. Having thus measured the electric tension of the air, and having lowered the conductor, I next place the wire, i, in communication with the electroscope, B, and by again raising the conductor, I ascertain whether the electricity be positive or negative. It is scarcely necessary to say that the conductor, when raised, gives electricity of the same nature as that prevailing at the moment in the atmosphere; and when lowered, manifests the opposite. In some conjunctures we must keep the conductor raised and in communication with the electroscope, in order to observe certain phenomena which I shall presently describe: this method I call observation with a *fixed conductor*.

I have also constructed a similar but portable apparatus for use on eruptive cones, when required.

Having given this description of the apparatus, it remains for me to relate the results obtained, especially on the occasion of the last eruption of Vesuvius.

The Observatory is distant, in a direct line from the central crater of Vesuvius, 2,380 metres, so that, when the smoke is copious, it is properly situated for the study of electricity, particularly when the wind inclines the pine-tree cloud in the direction of the Observatory, as frequently happened on the last occasion.

With smoke alone, without ashes, we obtained strong tensions of positive electricity ; with ashes only, which sometimes fell while the smoke turned in the other direction, we had strong negative electricity : when the smoke inclined towards the Observatory, accompanied with ashes and lapilli, we had sometimes one kind of electricity, and sometimes the other, just as the smoke or the ashes predominated ; and often with a "fixed conductor" we obtained negative electricity, and with a " movable conductor " positive electricity. In Naples, too, at the Meteorological Observatory attached to the University, my colleague, Professor Eugenio Semmola, observed negative electricity of strong tension whilst ashes were falling there in abundance. The tensions on this occasion were so strong as to equal those obtained at changes of weather or during storms (temporali), and, being beyond measure with a delicate electrometer, we marked them with the symbol ∞ : the same phenomena were observed when lightnings flashed.

When there is but little smoke, it is necessary to approach the eruptive mouths with a portable apparatus, in order to observe those phenomena which, in great eruptions, may be studied from the Observatory itself.

The conditions under which (folgori) lightning flashes are seen from the cloud of smoke are, that it is conveying great abundance of ashes. In 1861, there were small flashes even from the line of eccentric mouths

K 2

above Torre del Greco, although the smoke was not very great; and when these ceased to discharge, and the central crater became somewhat active, with a moderate amount of smoke but a great deal of ashes, small and frequent lightning flashes were observed in the twilight darting through the smoke, which was dark in colour. In 1850 the eruption was more vigorous, the smoke more abundant, and the ashes scarce, but the flashes were very rare. In 1855, 1858, and 1868, with a scanty supply of ashes and at intervals, no flashes were observed, and the electricity remained constantly positive. But having regard to the facts of antecedent eruptions, one sees that the flashes are always derived, from the midst of smoke accompanied with ashes and lapilli, which separate like rain from the rolling volumes of smoke, in the midst of which they were ejected.

But how can we account for the positive electricity of the smoke, and the negative electricity of the falling ashes? Without denying the probability that a part of the positive electricity depends upon the elevation of the smoke, as in the case of every other conductor we raise aloft, or with a jet of water sent from a vessel by compressed air, I think that the greater part of the electricity proceeds from the rapid condensation of vapours, which are changed from the gaseous condition into dense clouds; for even when the smoke issues tranquilly and does not rise, because carried away horizontally by the wind, it gives signs of positive electricity. From all my studies of atmospheric electricity, and from some experiments made specially, it follows that the condensation of vapours is the origin of this development of positive electricity.

The negative electricity of the falling ashes certainly arises from the fact itself of their fall; for if we place a metallic vessel full of ashes upon an elevated and well-situated terrace, while the atmospheric electricity is positive, and cause the ashes from the vessel to fall gradually into an insulated metallic cup, communicating with Bohnenberger's electroscope placed at three or four metres distance from the vessel, the electroscope will manifest negative electricity. If the upper vessel be insulated, and the ashes permitted to fall upon the ground, we shall obtain, from the vessel, positive electricity. The intensity of these electric manifestations depends (other things being equal) upon that predominant at the moment in the air; so that if the experiment be made while negative electricity prevails, the falling ashes will manifest positive electricity, the upper vessel then showing negative electricity. Now, as the ashes separate from the positively electrified smoke in order to approach the ground, which is negatively electrified, it follows that they must manifest negative electricity upon touching the ground, leaving the positive electricity in the smoke above. For this reason, the electric tension of the smoke is increased by the descent of the ashes and lapilli, so that discharges between the upper and lower part of the pine-tree cloud, or the surface of the crater, are rendered possible. Hence it follows that the flashes of lightning of Vesuvius play through the smoke, and with difficulty strike bodies upon the earth; and from this circumstance our ancestors believed the thunderbolts of Vesuvius to be harmless. However, if the smoke were very great, and driven by the force of the wind to some distance from the crater, with an abundant fall of ashes, it would be possible to have lightning flashes proceed from the smoke to the earth. I possess some documents which relate that, in 1631, thunderbolts fell upon the Church of Santa Maria del Arco, and other places on the coast of Sorrento.

After upwards of twenty years' study and observation of meteoric electricity, I am enabled to prove that atmospheric electricity is never manifested without rain, hail cr snow, and that manifestations of light are always accompanied by thunder—manifestations of light (*lampi*), thunder and rain being most closely connected. We may have rain without manifestations of light, but never the latter without rain or hail. I cannot here repeat what I have demonstrated in other memoirs; I can only say that the lightnings of Vesuvius, erroneously believed to be not accompanied by thunder, are really not accompanied by rain, but are induced by the descent of ashes and lapilli.(6)

GENERAL CONCLUSIONS.

WE may conclude from what I have stated:

1. That by the assiduous study of the central crater, and the indications afforded by the "Apparatus of Variations" and the "Electro-Magnetic Seismograph," we can obtain precursory signals of eruptions; and that the other premonitory signs pointed out by our ancestors, such as the drying up of wells, either only happen occasionally or are mere coincidences, such as those of the coincidence of a dry or a rainy season, the prevalence of certain winds, etc.*

2. That the fumaroles of the lavas are communications between the external surface of the lava, hardened and more or less cooled, and the interior lava still pasty, or at least incandescent.

---

* Earthquakes, though in distant regions, usually precede eruptions. The Earthquake of Melfi preceded the great Eruption of Etna in 1852; the Earthquake of Basilicata of December, 1857, terminated with the Eruption of 1858, which filled the Fossa Grande with lava; the Earthquakes of Calabria of 1867 and 1870 were the precursors of the Vesuvian conflagrations of 1868, 1871, 1872. A Volcano, also, in the Island of Java had a great eruption in the month of April, some days before the last conflagration of Vesuvius, as I learnt from a letter addressed to Signor Herzel, Swiss Consul at Palermo, communicated to me (?) by the astronomer, Signor Cacciatore.—*Palmieri.*

3. That from the lava, while flowing, there is no escape of acid vapours, neither from the fumaroles at the first period of their existence, but these, if they last long enough, arrive at an acid period.

4. That hydrochloric is the first acid that appears, combined afterwards with sulphurous acid, and, still later, with sulphuretted hydrogen.

5. That vigorous lava streams may have eruptive fumaroles. (See Translator's Note 2 to p. 94.)

6. That the sublimations follow a certain order in their appearance.   In the neutral period we get sea-salt mixed with some metallic oxides, the first of which is oxide of copper.   But in the great lavas, chloride of iron appears simultaneously with the acid period.   Hydrochloric acid transforms the oxides into chlorides, which, in their turn, change into sulphurets or sulphates on the appearance of sulphurous acid.

7. That the acids, by attacking the scoriæ, create new chlorides and sulphates, which are thus not products merely of sublimation.

8. That micaceous peroxide of iron—so common and abundant near the eruptive mouths—is very scarce and rare on the lavas, unless conveyed there from the craters.

9. That chloride of iron—so manifest on the fumaroles of the great lavas—is only found in small eruptions close to the discharging mouths.

10. That the frequency of chloride of iron in the lavas of great eruptions masks the order of transformation of the other products.

11. The fumaroles at the summit of Vesuvius present even greater gradations, for they often emit carbonic acid or pure watery vapour.

12. Lead, which I first discovered in the fumaroles of the lavas of 1855, is a constant product of fumaroles which have a certain duration. It is often obtained as a distinct and crystallized chloride, and often is found in combination with other products.

13. Oxide of copper is also a constant and primary (*primitivo*) product of fumaroles. The chloride and sulphate of copper are formed from the oxide, directly contrary to general belief.

14. I do not think that the chloride of calcium, which I found on this occasion in almost all the deliquescent sublimations, is a product peculiar to this eruption only, in which alone, however, I found it. I was, therefore, induced to look for it in other sublimates, in which I might possibly have overlooked it, as, without doubt, my predecessors have done, owing to the deliquescence of the chloride of iron with which it was constantly combined. I think that this chloride, in accordance with the general law, is transformed into a sulphate—a transformation which readily occurs on Vesuvius.

15. Copious and well-crystallized sal ammoniac is only found on the fumaroles of those lavas which have covered cultivated or wooded ground.

16. The scarcity of oxygen in the gases of fumaroles may possibly arise from the formation of the oxides which precede the chlorides.

17. Lavas give a continuous spectrum, although

covered with smoke, when looked at with Hoffmann's
spectroscope with direct vision.*

18. The smoke gives positive electricity, and the
falling ashes negative electricity.

---

* I have made a large collection of sublimates, which I purpose
examining with the spectroscope, and I shall be able to place some at
the disposal of experimentalists who may desire to pursue investigations of this kind.

REFERENCE TO THE PLATES.

NOTES

BY THE TRANSLATOR.

1 (P. 82, text). Professor Palmieri has not given any description in this Memoir of his seismograph—the instruments described being those only which have relation to atmospheric electricity. The following brief account of his seismograph will, therefore, form a not unsuitable complement to his Memoir. The instrument, in general terms, is of that class in which the wave movements are indicated by the displacement, relative or absolute, of columns of mercury in glass tubes. It is a self-recording instrument, composed of two distinct portions—one for record of horizontal, or rather of what are called undulatory shocks ; the other for vertical shocks. In point of general principle, therefore, it is very similar to that proposed by me ("Transactions, Royal Irish Academy," in 1846), and in certain respects appears to me less advantageous than the latter. Some account of the Palmieri instrument, together with some critical remarks as to its action, may be found in my "Fourth Report on Earthquakes" ("Reports, British Association, 1858," pp. 75–81). The following description of the instrument is derived from "The Engineer," of 7th June, 1872, and the publishers have to thank the proprietors of that journal for permission to use the illustration, Plate 8.

In Fig. 1, E is a helix of brass wire (gauge about one millimetre); the helix consists of fourteen or fifteen turns, and has a diameter of from twenty to twenty-five millimetres ; it hangs from a fine metal spring, and can be raised or lowered by a thumb screw. From the lower end of the helix hangs a copper cone with a platinum point ; the latter is kept close to the surface of mercury in the iron basin, f, which rests on an insulating column of wood or marble, G. The distance of the point from the surface of the mercury remains constant, as the metal pillar, T, is of such a length that its expansion or contraction by change of temperature compensates that of the helix ; the latter is in connection (by T) with one pole of a Daniell's battery of two cells, and the basin, f, is connected with the other pole. Any vertical movement, however slight, makes the platinum point dip into the mercury, and thus completes the circuit. In this circuit are included two electro-magnets, C and D ; these, during the circulation of a current, attract their armatures, which are connected with levers. The action of C's lever is to stop the clock, A, which thus records, to a half-second, the time of the occurrence of the shock, at the same instant that the clock strikes an alarm bell, which attracts the attention of an observer. The lever, attached to the armature of D, at the first instant of the current frees the pendulum of the clock, B, which was before kept from swinging, in a position out of the vertical ; the clock then acts as a time-piece, and its motion unrolls a band of paper, $k \, k \, k$, at a rate of three metres an hour. At the same time the armature of D, while attracted, presses a pencil point against the band of paper which passes over the roller, m, marking on it, while the earthquake lasts, a series of points or strokes which occupy a length of paper corresponding to its duration, and which record the work of the shock. After it is over the paper continues to unroll from the drum, i, and passing round the clock, rolls on to the drum, l. If a fresh shock occur the pencil indicates it, as before, on the paper, and the length of blank paper between the two sets of marks is a measure of the interval of time between the shocks. By way of additional check, several helices, $h \, h \, h$, are hung from a stand, with small permanent magnets suspended from their ends ; below and close to these latter are small basins, holding iron filings ; into these the points of the magnets dip, when their helices oscillate vertically, and some filings remain sticking to the magnets as a record of the shock. One of the magnets has a shoulder on it which moves an index hand along a graduated arc, as shown in Fig. 2, thus again registering the amount of the vertical movement. Such are the arrangements intended for the record of the undulatory or horizontal elements of the wave of shock.

The following are the arrangements proposed for recording the horizontal motions : On the stand, to the right of the clock, A, are set four U-shaped glass tubes, open at their ends. One of each pair of vertical branches must have a diameter at least double that of the other. These pairs, with their supporting columns, are shown in plan, where one pair lies N. and S., another E. and W., a third N.E. and S.W., and the other N.W. and S.E. It will be observed that metallic bars pass from the pillar, P, over the ends of all the long branches, and similar bars pass from R, over the ends of the short branches ; the pillars themselves, as in the case of the other instruments, are each connected with one pole of a Daniell's battery, the connections including the electro-magnets, C and D. The description of one U tube, n, will apply to all the others ; n is partly filled with mercury, and an iron or platinum wire, o, suspended from the bar above the short branch, dips into the mercury therein, while another platinum wire hung from the bar over the mouth of the longer branch, has its end very close to the surface of the mercury in that branch. Any shock which is not perpendicular in direction to the plane of the branches of the U will cause the mercury to oscillate in the tubes, and more sensibly in that with the smaller diameter ; when it rises up in

the latter, so as to touch the platinum point, the connection between
P and R is made and the circuit completed, starting the action of the
electro-magnets C and D, which record the shock, as already described.
By having the planes of the tubes set in the different azismuths, already
mentioned, one or more of the pairs is sure to be acted upon, and by
observing in which the oscillation takes place the direction of the
shock is supposed to be ascertained. Besides this, each long branch
of the U, viz., that of smaller diameter, has a small ivory pulley, q,
fixed above it, over which passes a single fibre of silk, with an iron
float at one end, resting on the surface of the mercury ; at the other
end of the fibre hangs a counterpoise ; fixed to the pulley is a fine
index hand, capable of moving along a graduated arc. When the
shock takes place the mercury, rising in the long branch, raises the
float on its surface, the silk fibre at the same time makes the pulley
revolve with its index hand, which afterwards remains stationary,
as the counterpoise prevents the float from sinking again with the
mercury. The reading on the graduated arc is thus a measure of the
movements produced in the instrument by the horizontal element of
the shock, and is supposed to measure that shock. It is assumed
that in all these instruments shocks, however small, can be recorded
with certainty by adjusting the distance between the platinum points
and the mercury.

The arrangement of Daniell's battery used for the seismograph is
shown in Fig. 4, where, for convenience of cleaning, the copper element is made of wire (about No. 8 Birmingham wire gauge) coiled
flat without the spirals touching. Crystals of sulphate of copper are
placed at the bottom of the outer cell, into which water is poured ;
and the inner cell, into which the zinc plate goes, is filled with sili-
cious sand.

In addition to the above some instruments of a rougher description
are employed as checks. Thus, at the foot of the pillar, G, there is a
wooden trough with eight holes, facing as many equidistant points of
the compass (two of them shown in section) round its inner circum-
ference ; mercury is poured into the basin until its level is nearly up to
the lips of the holes. The effect of a shock is to throw some of the
mercury into one or more of these holes, and the greater the oscillation the more mercury is thrown into the cells through the holes.
The screws shown outside are for drawing off the mercury from the
cells, when its quantity can be measured. The direction of the shock
is shown by seeing which cells are filled with mercury. This is the
old Cacciatore seismometer which has been long employed in Italy.
(See 4 " Report of British Association, 1858," p. 73), and Daubeny's
" Volcanoes," Appendix. The following is another contrivance. From
the arm of the pillar, G, a fine metal wire hangs, with a metal ball at
its end, which, by its oscillation, thrusts out one or more light glass
tubes, set horizontally in a stand, as shown in Fig. 3. The two rings
are of wood, and the glass tubes pass through holes in them ; small
leather washers are placed outside the outer rings : the displacement

of one or more tubes is assumed to measure the horizontal element of the shock. By means of this apparatus the time of the first shock is recorded, as well as the interval between the shocks, and the duration of each ; their direction, whether vertical or horizontal, is given, as also the maximum of intensity. Professor Palmieri has the instruments examined three times a day, and an assistant-observer is always at hand to attend to the bell, and put back the apparatus to its normal position for fresh observation.

It has been stated that this instrument is sensible to most of the shocks which occur in the Mediterranean basin.

It is not my intention here to offer any criticism as to the construction or performances of this instrument, the rather as I must confess I do not quite share the high opinion of its inventor as to the certainty or exactitude of its indications.

There can be no question as to the extreme importance to science of the establishment and continued use of a seismographic instrument of unexceptionable construction at the Observatory upon Vesuvius; and it would be a valuable gift to science, were the Italian Government to enable Signor Palmieri to establish such an one. Its great value and the very first problem to set the instrument to solve should be, by *a rigid determination of the direction of propagation of the wave of shock*, of those slight or stronger pulsations which precede or accompany the Vesuvian like all other eruptions, on arriving at the Observatory, *to fix the depth, and the position vertically beneath the cone, whence these pulses are derived.* This would be, in fact, to fix the depth and position beneath the mountain at which the volcanic focus is situated for the time, or, at least, where the volcanic activity is at the time greatest. And the assured knowledge, even within moderate limits of accuracy, of this depth, and even for this single mountain, would be an immense accession to our positive knowledge, and a really new stage gained for future advances. At present, we know but little as to the actual depth below our globe's surface at which volcanic activity occurs, or to which it is limited, either upwards or downwards. I have, myself, established some data upon the flanks of Etna, not yet published, which may enable me to afford some information on the subject hereafter. Meanwhile, Professor Palmieri possesses unrivalled opportunities for such observations; and I trust health, life and means may be afforded him, to become the first who shall have made this great addition to our positive knowledge of Vulcanology.

So far, popularly at least, the alleged chief uses and value of these seismographic instruments, at the Observatory of Vesuvius, have been made to depend upon their being presumed to afford means for foretelling eruptions, or affording precursory warnings of their probable progress and destructive course.

I feel compelled to express my own total disbelief in the possibility of any such predictions in the present state of science, by the help of any instruments whatsoever, of such a nature as to be of any *prac-*

tical value, or any certainty beyond that which a certain amount of *mere experience* as to the *rôle commonly played* by Vesuvius or other Volcanoes in pretty habitual activity affords to the observer for a lengthened period. And even this affords scarcely any guide as to what may happen next. Monte Nuovo was thrown up in a night; Vesuvius *might* double its volume in a night, or might sink into a hollow like that of the Val del Bove in a not much longer time. A small *fusillade* may go on for months, and yet, without an hour's notice, by any premonitory sign, may waken up to a roar and darken the air with ashes and lapilli such as those which overwhelmed Pompeii. One eruption may blow forth little but dust and ashes (so called), another may pour out rivers of lava and little else.

The *main* mischief of all eruptions is effected in two ways: by the deposit of dust and ashes, lapilli, etc., to the injury or destruction of fertile land, and by the streams of lava which overwhelm it, as well as buildings, etc. But what information of any value can seismographic observation afford as to the course that either of these may take in any eruption? The volume of pulverulent material that may be ejected cannot be foreseen; its distribution depends mainly upon its nature and upon the direction and force of the wind at the time; or again, how shall these warn us as to the course that the lava, if it appear, shall take, when we cannot possibly foretell when, how, or by what mouth it may issue. Even in this late eruption of 1872, with Palmieri stoutly at his post upon the mountain, and the Observatory instruments in full activity, they gave no forewarning of the sudden and unexpected belch forth from the base of the cone, of that tremendous gush of liquid lava which in a few minutes cut off from life the unhappy visitors whose deaths he has recorded.

———————

2 (P. 94). It can scarcely be supposed that these small eruptive-looking belchings forth from the lava stream, *en route*, are truly of an eruptive nature at all, *i.e.*, in any way connected with forces seated deeply beneath the bed of the lava stream, or in any way connected with the volcanic ducts of the cone or beneath it. They are most probably merely the bursting upwards of large bubbles; that is, of cavities formed in the mass of the more or less liquid lava by intestine movements, as its mass winds and rolls along, and by the aggregation of smaller cavities—all being filled with steam and gases—together with dust and volatile products which are ejected when the cavity opens up, and its contents escape at the upper surface of the lava stream in virtue of the continuation of the twistings and convolutions due to the stream motion itself, and to the unbalanced hydrostatic pressures acting upon the parietes of the bubble. Very large single bubbles of like character rise in the fluid lava within craters in vigorous action, and often so regularly that their recurrence causes a sort of rhythmical rush and roar in the column of steam, etc.,

L

issuing above the mouth. This was evident in the discharges issuing in 1857 from the highly instructive minor *bocca*, then existing, examined by me, and referred to ("Report, Naples, Earthquakes," etc. Vol. II., pp. 313, 314), as presenting at the time great facilities for determining pyrometrically the temperature of the lava within, and of the dry superheated steam issuing with a rhythmic roar from it. M. Le Coq ("Époques Géologiques d'Auvergne," Tome IV.) has recorded some examples of the formation and opening-out of large bubble-like cavities in lava already ejected. Perhaps that able and laborious vulcanologist, whose death a few months ago science still deplores, attributes too much importance as well as magnitude to them, when attributing the formation of what he has denominated "craters of explosion," to the mechanism of the rise and bursting of such bubbles upon a gigantic scale. Such blowings forth, sudden or prolonged, from particular spots of lava streams, *en route*, undoubtedly may also have their origin in damp places, or water or air-filled cavities in or beneath the bed over which the lava rolls, which, getting gradually heated, generate steam, or air or gases under tension by expansion, etc., which thus at length blow through the liquid or pasty lava flowing above, and which in bursting through delivers much dust also, and so simulates a little eruptive crater. Examples of this, upon a great and convincing scale, can be pointed to in the Val di Calanna and elsewhere on Etna.

---

3 (P. 96). There are strong grounds for the gravest doubts that there exists any real connection of a physical character between Volcanic Eruptions, and Earthquakes more or less *approximately* coincident only, in time of occurrence ; the respective sites being widely apart, and the less the probability as the intervening distance is greater. The discussions of the large number of records that are to be found of such coincidences—mostly but partial, and in but *very* few instances complete coincidences—by Perrey, von Hoff, and others, as well as by myself, do not tend to sustain the view that such imperfect contemporaneity is based upon any causative connection. The seismic region of Greece appears to have no *direct* connection with that of Southern Italy : the band of connection, if any, seems to lie between Northern Italy, across the Northern Adriatic, by Ragusa, and thence spreading into Asia Minor.

---

4 (P. 97). The abundance of coleoptera and of various other forms of insect life about lava beds, both recent and old, is a very singular fact, and one worthy of the careful observation of entomologists. In the autumn of 1864, at mid-day, when sitting sketching upon the lava about the middle of the Val del Bove (Etna), I found it almost impossible to work, or even to remain for an instant still, in consequence of the continual cloud of insects, large and small, that struck against me in flight, endangered the eyes, and swarmed upon

my clothes. It is quite possible that this local superabundance of insect life may arise merely from the general dryness and warmth of such places, and the plentiful *nidus* that the innumerable cavities in lava afford for the eggs and earlier stages of insect life ; still, this apparition of one form of life may also be connected with other circumstances not unimportant to discover.

5 (P. 120). The *Crocella* is a small wooden cross, erected several years ago, and which one passes to the right hand at the upper end of the path along the ridge of tufa and volcanic conglomerate upon which the Observatory stands, in ascending thence to the Atria del Cavallo.

6 (P. 134). That the causes assigned by Professor Palmieri for the potent developments of electricity (positive or negative) which characterise the ascent of the issuing columns of (chiefly if not always) *dry* steam, with a relatively small volume of various gases, and throwing up, in their blast, volumes of small solid particles in ashes and lapilli, etc., and the subsequent fall as a mineral or stony hail-shower of the latter, through the partially condensing vapours and the circumambient air, are the main causes of electrical development evidencing itself in lightning flashes, is no doubt true. We must not, however, lose sight of the many other and very effective agencies at work here to produce electric excitement. The actual *bocca* of the volcanic vent whence the steam roars off constitute the cone a veritable hydro-electric machine. Mechanical energy in various forms is transformed into electric energy. Chemical action is going on both in the solid and in the vapourous and gaseous emanations as they rush into and remain in the air or descend from it, and chemical action is transformed in part into electric energy. Percussion between ascending and descending particles and fragments, fractures and breaking up of more or less of these, thus and by sudden changes of temperature in cooling, are likewise operative. In addition, great and violent movements in the atmosphere itself result from the large local accessions of temperature by the heated volume driven up into it, and which in turn give rise to electric disturbance of the same character as those produced in wind storms and whirlwinds, brought about by the natural causes which every day effect disturbances in our atmosphere all over the globe.

7 (P. 135). The views stated in note 3 (to page 96) may here again be referred to as in point. How is it possible, in the present state of science at least, to establish any physical connection between an eruption in Java and one of Vesuvius, "with half the world between," when not even having the solitary connecting link of complete contempo-

raneity, and which, if it existed, yet might be nothing but accidental? A list of shocks upon record, which have occurred more or less nearly simultaneously at distant parts of the world, may be found in my fourth Report, (" Facts of Earthquakes," "British Association Reports, 1858 ") and the reasons are there given for rejecting the notion of any direct physical connection between the origins of the respective shocks.

Shocks, emanating from the close neighbourhood of volcanic vents, or simultaneity of eruption, in vents not far distant from each other, stand upon a different footing.

PLATE IA.

PLATE IIA.

PLATE IIIA.

PLATE IV A.

PLATE Vₐ.

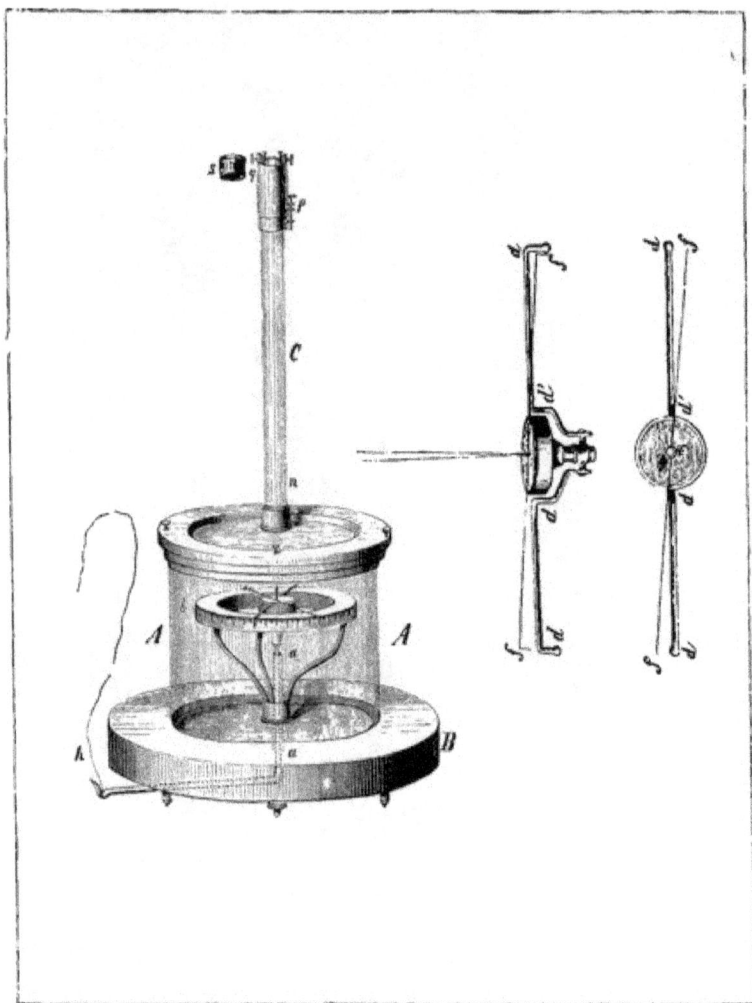

PLATE VIA

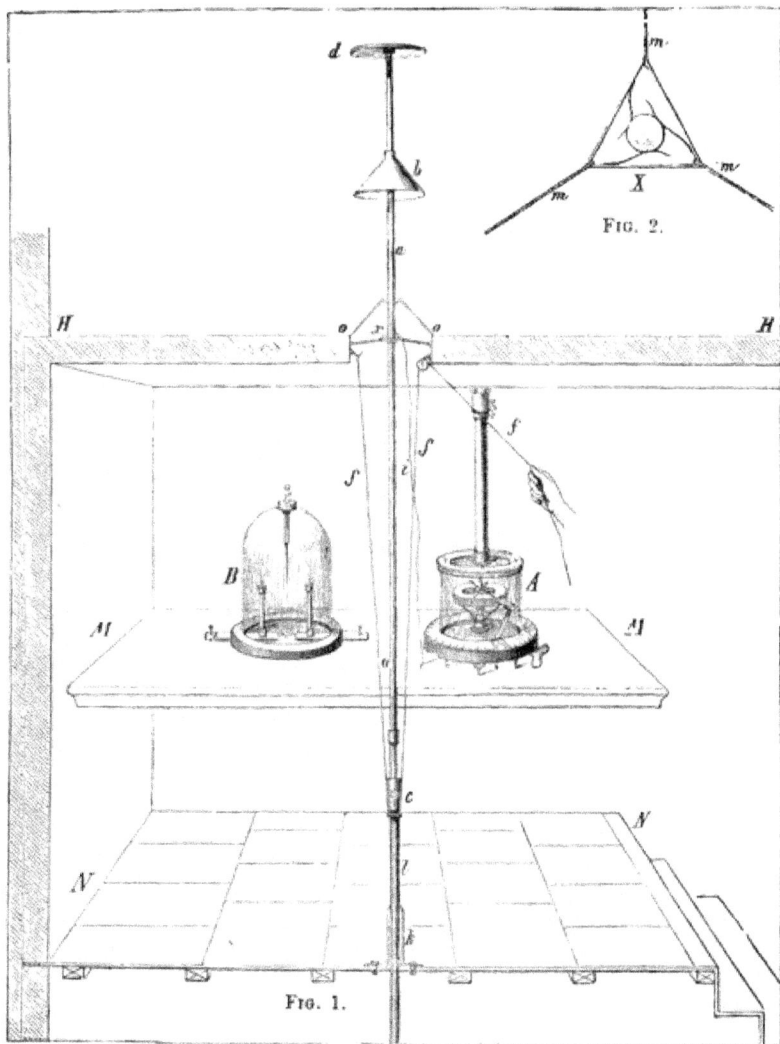

FIG. 2.

FIG. 1.

PLATE VIIA.